Surface Science of Degradation and Surface Protection

Surface Science of Degradation and Surface Protection

Guest Editors

Matic Jovičević-Klug
Patricia Jovičević-Klug
László Tóth

MDPI

Basel • Beijing • Wuhan • Barcelona • Belgrade • Novi Sad • Cluj • Manchester

Guest Editors
Matic Jovičević-Klug
Sustainable Magnets and Recycling Group
Max Planck Institute for Sustainable Materials
Düsseldorf
Germany

Patricia Jovičević-Klug
Department of Interface Chemistry and Surface Engineering
Max Planck Institute for Sustainable Materials
Düsseldorf
Germany

László Tóth
Bánki Donát Faculty of Mechanical and Safety Engineering
Óbuda University
Budapest
Hungary

Editorial Office
MDPI AG
Grosspeteranlage 5
4052 Basel, Switzerland

This is a reprint of the Special Issue, published open access by the journal *Coatings* (ISSN 2079-6412), freely accessible at: https://www.mdpi.com/journal/coatings/special_issues/Surface_Protection_Degradation.

For citation purposes, cite each article independently as indicated on the article page online and as indicated below:

Lastname, A.A.; Lastname, B.B. Article Title. *Journal Name* **Year**, *Volume Number*, Page Range.

ISBN 978-3-7258-8381-3 (Hbk)
ISBN 978-3-7258-8382-0 (PDF)
https://doi.org/10.3390/books978-3-7258-8382-0

Contents

coatings

MDPI

icle

ptimization and Mechanism Study of Bonding Properties of FRP/Al7075 Single-Lap Joints by Low-Temperature Plasma urface Treatment

vei Wen *, Ruozhou Wang and Entao Xu

College of Material Science and Technology, Nanjing University of Aeronautics and Astronautics, Nanjing 210016, China; wangruozhou@szcu.edu.cn (R.W.); xet980624@nuaa.edu.cn (E.X.)
* Correspondence: wenliwei@nuaa.edu.cn

Abstract: This paper studied favorable low-temperature plasma (LTP) surface treatment modes for Carbon Fiber Reinforced Polymer (CFRP)/Al7075 single-lap joints using complex experimental methods and analyzed the failure modes of the joints. The surface physicochemical properties of CFRP after LTP surface treatment were characterized using scanning electron microscopy (SEM), contact angle tests, and X-ray photoelectron spectroscopy (XPS). The influence mechanism of LTP surface treatment on the bonding properties of CFRP/Al7075 single-lap Joint was studied. The results of the complex experiment and range analysis showed that the favorable LTP surface treatment parameters were a speed of 10 mm/s, a distance of 10 mm, and three repeat scans. At these parameters, the shear strength of the joints reached 30.76 MPa, a 102.8% improvement compared to the untreated group. The failure mode of the joints shifted from interface failure to substrate failure. After low-temperature plasma surface treatment with favorable parameters, the CFRP surface exhibited gully like textures, which enhanced the mechanical interlocking between the CFRP surface and the adhesive. Additionally, the surface free energy of CFRP significantly increased, reaching a maximum of 78.77 mJ/m^2. XPS results demonstrated that the low-temperature plasma surface treatment led to a significant increase in the content of oxygen-containing functional groups, such as C-O, C=O, and O-C=O, on the CFRP surface.

Keywords: carbon fiber reinforced polymer; low-temperature plasma surface treatment; single-lap joint; shear strength; surface physicochemical properties

1. Introduction

Carbon fiber is a fiber material composed of carbon atoms. It possesses not only low density and high strength but also excellent high-temperature and low-temperature resistance, good thermal shock resistance, a low thermal expansion coefficient, and outstanding corrosion resistance. Carbon fiber reinforced polymers (CFRP) find extensive applications in critical industries such as aerospace, automotive, rail transportation, wind energy, and construction [1].

The use of CFRP is the primary approach for aircraft lightweighting in the current aviation industry. The range and application areas of CFRP have become crucial indicators for assessing the level of aircraft structural advancement. Additionally, they serve as important criteria for airlines when purchasing aircraft. With the continuous increase in the proportion of CFRP in aircraft components, the issue of connecting CFRP with traditional lightweight alloys such as aluminum alloys has become inevitable. The traditional mechanical joining method has advantages in terms of load transmission and ease of disassembly, making it the main joining method for aircraft structures [2]. When mechanical joining is employed between composite materials and metals, the differences in their mechanical properties often lead to reduced connection efficiency. Moreover, joint strength, stiffness, and durability are compromised. Compared to traditional mechanical joining methods, adhesive bonding offers advantages such as lightweight structures and stress-free concentration, leading to its increasingly widespread application in composite connections [3]. Adhesive bonding techniques are commonly used for connecting secondary load-bearing components or participating in connections in an auxiliary manner [4–6]. Adhesive bonding techniques complement mechanical joining by providing uniform stress distribution to joints, thereby avoiding localized stress concentration [7,8]. Therefore, enhancing the bonding properties of adhesive joints has become an important issue in the practical application of composites.

Due to the low surface energy and high chemical inertness of CFRP surfaces, when CFRP is directly involved in adhesive bonding, the interfacial bond strength between CFRP and the adhesive layer is extremely low, making the joint highly susceptible to failure.

Therefore, it is necessary to perform certain surface treatments to improve the bonding strength of CFRP. Common surface treatment methods used before bonding include solvent cleaning, mechanical surface treatment, chemical surface treatment, plasma treatment, laser treatment, and so on [9,10]. Among these methods, mechanical surface treatment can cause significant damage to the substrate surface and generate contaminants. Chemical surface treatment can lead to environmental issues due to chemical emissions and low efficiency. Laser treatment carries the risk of fiber damage and delamination [11]. Low-temperature plasma (LTP) surface treatment, as a dry surface modification technique, can introduce reactive functional groups on the surface of CFRP, thereby increasing its surface energy and effectively improving its bonding properties.

Compared to other surface treatment methods, low-temperature plasma surface treatment also offers advantages such as cleanliness and high efficiency. It holds great promise in the field of composites due to these advantages [12–17].

In practical applications, artificial gas discharge methods are commonly used to generate stable and controllable low-temperature plasmas. Some commonly used gas discharge methods include glow discharge, dielectric barrier discharge, corona discharge, and jet discharge [18,19].

The academic community has extensively researched the application of low-temperature plasma treatment for surface modification of composites. Chang et al. [20] investigated the effects of acetone cleaning, plasma treatment, and sandblasting treatment on the bonding properties of CFRP adhesive joints. The results indicated that both plasma treatment and sandblasting improved joint strength. However, plasma treatment primarily led to cohesive failure of the adhesive, while sandblasting caused interfacial damage to the CFRP, indicating fiber damage during the sandblasting process. In terms of improving lap shear strength, plasma treatment was found to be more effective than sandblasting. Matthias et al. [21] used low-temperature atmospheric pressure plasma generated by a plasma jet to improve the bonding properties of polypropylene (PP). They conducted a comprehensive surface analysis by employing various O_2/N_2 mixtures. The results showed that the majority of plasma-induced polar functional groups were present in the form of water-soluble low molecular weight oxide species, within a few nanometers of the surface, regardless of the composition of the gas mixture. These chemical changes also indicate that there are differences in the interaction mechanisms between plasma reactants and the polymer surface, resulting in variations in the bonding properties of the polymer. Rafailovic et al. [22] employed atmospheric pressure plasma treatment to activate CFRP surfaces and enhance the properties of electroplated copper coatings on them. The results showed that a sufficiently long treatment time resulted in significant etching of the top epoxy resin layer of the carbon fibers, making them more conducive to subsequent copper metallization. The bonding properties of the Cu-carbon fiber composite layer were enhanced through longer plasma activation, eliminating the need for any additional chemical surface treatment apart from the standard solution used for self-catalytic Cu deposition.

Chris et al. [23] investigated the impact of atmospheric plasma treatment (APT) on the surface properties of amine-cured carbon fiber/epoxy resin composites. The results indicated that APT significantly increased the surface roughness of the solvent-wiped surface, and the silicon-containing substances reacted with oxygen species in the plasma, forming a layer of silicon dioxide. APT slightly reduced the shear strength of the solvent-wiped surface in lap shear samples, while it increased the shear strength of samples contaminated with organic silicon.

Based on the current state of research, the majority of studies have focused on comparing low-temperature plasma surface treatment with other surface treatment methods, highlighting its advantages in improving the strength of composite material adhesive joints. However, in practical applications, many variables are involved in the low-temperature plasma surface treatment process, such as treatment velocity, treatment distance, number of repeat scans, discharge power, gas composition, etc. Surprisingly, there is limited research on the coupling effects of these variables on the strength of adhesive joints. Therefore, this study employed a complex experimental design to investigate the effects of three parameters—velocity, distance, and number of repeat scans—on the bonding properties of CFRP/Al7075 single-lap joints during the LTP surface treatment process. By utilizing the range analysis method, the main parameters influencing the bonding strength of the joints were identified. Next, the favorable parameters from the complex experiment were selected to further optimize the main influential parameters. This resulted in the determination of the LTP surface treatment parameters that achieved the highest bonding strength of the joints. Additionally, a characterization analysis was conducted to examine the macroscopic failure modes of the joints after LTP surface treatment. Finally, a characterization analysis was conducted on the surface wettability, surface topography, and surface chemical structure of CFRP after LTP surface treatment, aiming to investigate the mechanism of LTP surface treatment on the bonding properties of CFRP/Al7075 single-lap joints.

2. Materials and Methods

2.1. Experimental Materials

The CFRP (EH918, Hengshen, Zhenjiang, China) used for preparing the lap joints was manufactured by curing EH918 epoxy resin with HF40C carbon fiber(Hengshen, Zhenjiang, China). The properties of EH918 epoxy resin and HF40C carbon fiber are shown in Tables 1 and 2.

Table 1. Properties of EH918 epoxy resin.

Density	Tensile Strength	Flexural Strength	Elongation at Break	Curing Temperature
1.26 g/cm^3	105 MPa	80 MPa	23%~26%	210 °C

Table 2. Properties of HF40C carbon fiber.

Linear Density	Tensile Strength	Tensile Modulus	Elongation at Break	Density
445 g/km	5600 MPa	295 GPa	1.9%	1.80 g/cm^3

The aluminum alloy used for preparing single-lap joints was 7075-T6 high-strength aerospace aluminum alloy (Jingteng Metal, Shanghai, China) and its properties are shown in Table 3.

Table 3. Properties of Al7075-T6.

Density	Yield Strength (0.2%)	Tensile Strength	Shear Strength	Elastic Modulus	Hardness
2.81 g/cm^3	503 MPa	572 MPa	330 MPa	71.7 GPa	150 HB

The adhesive used for preparing single-lap joints was DP460, a high-strength two-component epoxy structural adhesive (DP460, Minnesota Mining and Manufacturing Corporation, Saint Paul, MN, USA). The main properties of DP460 are shown in Table 4.

Table 4. Properties of DP460 adhesive.

Elastic Modulus	Tensile Strength	Shear Strength	Elongation at Break
2.7 GPa	37 MPa	32 MPa	4%

2.2. LTP Surface Treatment Device

The LTP surface treatment device used in this study was a pulsed corona discharge low-temperature plasma jet generation device. It mainly consists of three components: a pulsed high-voltage power supply, a vortex blower, and a plasma spray gun, as shown in Figure 1.

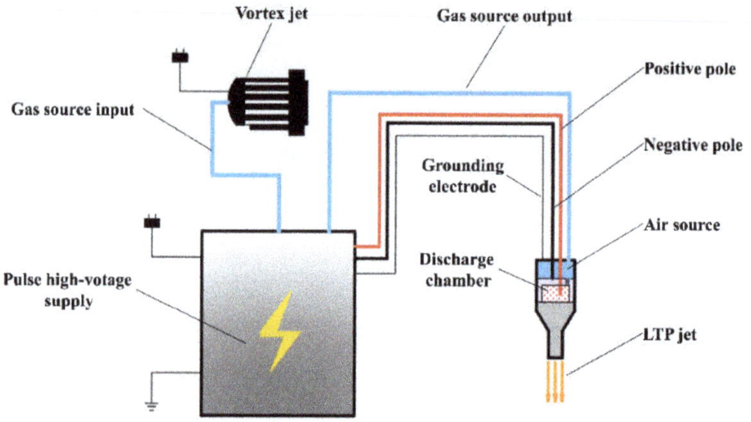

Figure 1. LTP surface treatment device.

The basic principle of the LTP surface treatment device is as follows: when the pulsed high-voltage power supply is activated, a high-frequency, high-voltage AC electric field is formed between the small curved positive electrode inside the discharge chamber and the surrounding negative electrode. This electric field causes the air inside the discharge chamber to dissociate, forming a sustained discharge of low-temperature plasma. Once the power of the LTP surface treatment device is stable, the vortex blower is activated to continuously supply compressed air into the discharge chamber. This airflow helps to expel the low-temperature plasma from the discharge chamber, forming a stable LTP jet. In this study, only air was used as a plasma-forming gas for LTP; no other gases were used.

In addition, this study combined the LTP jet generation device with a three-axis motion platform, enabling control over the velocity and distance of LTP surface treatment. The three-axis motion platform utilized a UMAC (Universal Motion and Automation Controller) motion control system (OMRON, Kyoto, Japan). This motion control system, paired with the Pewin32 Pro2 software, allowed for control over the motion parameters of the plasma spray gun in the X, Y, and Z directions.

2.3. Sample Preparation

According to GB/T 7124-2008, "Adhesives—Determination of tensile lap-shear strength of rigid-to-rigid bonded assemblies", CFRP and Al7075 were processed into standard dimensions of 100 mm × 25 mm × 2 mm. The standard adhesive bonding sample size is shown in Figure 2.

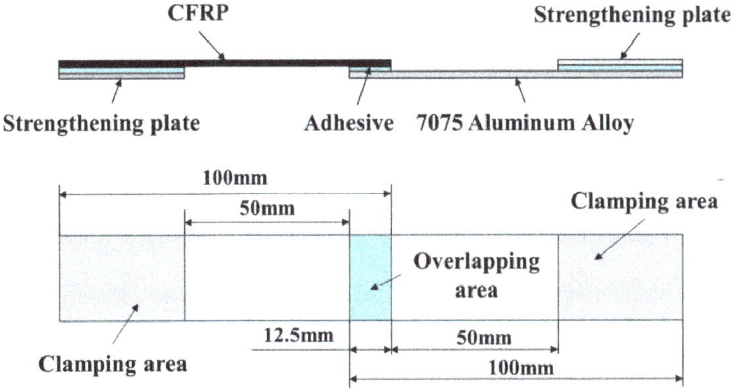

Figure 2. Standard adhesive sample size.

To eliminate the influence of other factors on the surface properties of the samples, it was necessary to perform certain pretreatments on the samples before conducting LTP surface treatment. According to GB/T 21526-2008 "Guidelines for Surface Treatment of Metals and Plastics Before Structural Adhesive Bonding", the CFRP and Al7075 surfaces were ultrasonically cleaned using acetone and dried in a vacuum oven. For Al7075, 800-grit water abrasive sandpaper was used for horizontal, vertical, and circumferential sanding to remove the weak oxide layer on the surface of Al7075 before acetone cleaning. Once the sample surfaces had been properly cleaned, aluminum strengthening plates with a thickness of 2 mm were attached to the clamping areas.

After the pretreatment of the samples was completed, the CFRP underwent LTP surface treatment using the LTP surface treatment device with different process parameters, as shown in Table 5. During the processing, scanning of the same surface area was repeated 1 to 3 times, depending on the test parameters. Five samples were prepared for each parameter group, and the LTP surface treatment process is shown in Figure 3.

Table 5. LTP surface treatment parameters.

Level	A	B	C
	Velocity (mm/s)	Distance (mm)	Number of Repeat Scans
1	5	10	1
2	10	15	2
3	15	20	3

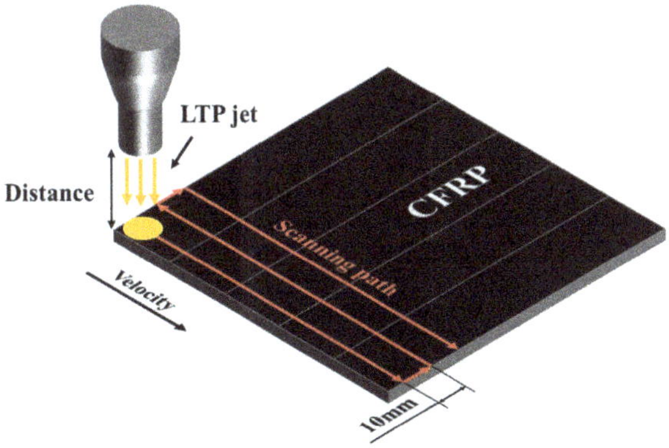

Figure 3. LTP surface treatment process.

After the samples underwent LTP surface treatment, the DP460 adhesive was dispensed into a 1:1 mixing nozzle using a dispensing gun, ensuring thorough and uniform mixing. The mixed adhesive was then applied onto the overlapping area of the sample. According to the testing standard GB/T 7124-2008, the typical adhesive layer thickness was 0.2 mm. To ensure consistency in the thickness of the adhesive layer, two parallel 0.2 mm metal wires were used along the loading direction for control purposes.

As shown in Figure 4, metal wires parallel to the load direction were placed on the surface of the sample. The end of the sample was then aligned with the 12.5 mm overlap area line. Fixtures were used to secure both sides of the overlapping area and apply equal pressure to the sample for curing. Finally, a tool was used to remove the excess adhesive. To eliminate errors caused by environmental temperature and humidity, each group of five samples was placed in a vacuum drying oven at 30 °C for 4 h to ensure complete curing of the adhesive. After the adhesive had fully cured, the samples were removed from the oven and subjected to testing.

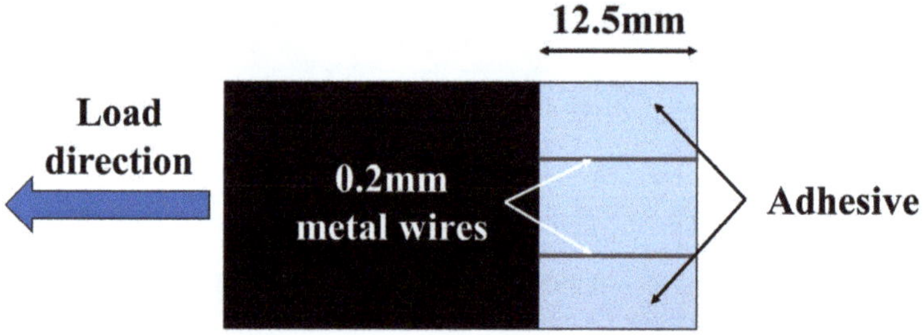

Figure 4. Placement of metal wires.

2.4. Shear Strength Test

The determination of shear strength for a single-lap joint follows the standard GB/T 7124-2008 titled "Determination of Shear Strength of Adhesives (Rigid-to-Rigid Materials)". The testing procedure was conducted using an electronic universal testing machine (CMT5150, Sans, Shenzhen, China). Throughout the testing process, the loading rate was maintained at 2 mm/s. When the specimen failed during testing, the peak load and force-displacement curve of the specimen were measured and recorded. The tests are repeated five times for each set of parameters, and the average values were calculated. The shear strength of the specimens was then determined using Equation (1).

$$\tau = \frac{F_m}{B \times L} \tag{1}$$

In Equation (1), τ represents the shear strength of the specimen in megapascals (MPa), F_m represents the peak load of the specimen in newtons (N), and B and L represent the lap width and lap length of the joint in millimeters (mm).

2.5. Characterization of Surface Physicochemical Properties

The failure modes of CFRP/Al7075 single-lap joints with different LTP surface treatment parameters were characterized using a Digital Vision Microscope (DVM-6V, Leica, Wetzlar, Germany). By observing the morphology of the joint fracture surfaces, the failure modes of joints under different processing parameters were analyzed to reveal the influence mechanism of low-temperature plasma surface treatment on the bonding strength of the joints.

The surface morphology of CFRP with different LTP surface treatment parameters was characterized using a scanning electron microscope (Sigma 300, ZEISS, Oberkochen, Germany). Before scanning, a sputter coater (Quorum SC7620, Oxford Instrument Technology, Ltd, Oxford, UK) was used to coat the sample surface with a thin gold layer for

45 s. The scanning process was performed at an accelerating voltage of 3 kV, with a probe current ranging from 3 pA to 20 nA. By observing the microstructure of the sample surface, the impact of low-temperature plasma treatment on the microstructure of CFRP surfaces was analyzed.

The surface chemical structure of CFRP with different LTP surface treatment parameters was characterized using an X-ray photoelectron spectrometer (Thermo Fisher Scientific, Waltham, MA, USA). The XPS instrument utilized Al Kα radiation as the excitation source (hv = 1486.6 eV). The analysis chamber maintained a vacuum level better than 5.0×10^{-7} mBar, and the working voltage was set at 12 kV. Before the experiment, the CFRP was cut into samples with dimensions of 10 mm $\times$ 10 mm $\times$ 2 mm, and the samples were cleaned using acetone to remove any contaminants from the surface. Then, the samples underwent low-temperature plasma surface treatment using different parameters. The full-spectrum scan step size was 1 eV, and the narrow-spectrum scan step size was 0.1 eV. By analyzing the XPS data, the effects of low-temperature plasma treatment on the chemical structure of the CFRP surface were investigated.

The contact angles of the surface of CFRP after LTP surface treatment were measured using a Contact Angle Measurement Instrument (JC2000D2S, Powereach, Shanghai, China). For each group of samples, the contact angles at different positions on the surface were measured five times, and the average value of the measurement data was taken. A droplet volume of 2 μL was used, and water and ethylene glycol were chosen as the test liquids. The surface free energy parameters of the two tested liquids are known, as shown in Table 6. According to the Owens-Wendt-Rabel-Kaelble method [24], the surface free energy of CFRP can be calculated from the joint Equations (2) and (3).

$$\gamma_L(1 + cos\theta) = 2\sqrt{\gamma_S^d \gamma_L^d} + 2\sqrt{\gamma_S^p \gamma_L^p} \quad (2)$$

$$\gamma_S = \gamma_S^d + \gamma_S^p \quad (3)$$

Table 6. Surface energy parameters of water and glycol [24].

Test Liquid	γ_L^d (mN/m)	γ_L^p (mN/m)	γ_L (mN/m)
Water	21.8	51.0	72.8
Glycol	29.3	19.0	48.3

In Equations (2) and (3), γ_L is the surface free energy of the test liquid, θ is the contact angle, γ_S^d and γ_S^p are the polar and dispersion components of the surface free energy of CFRP, γ_L^d and γ_L^p are the polar and dispersion components of the surface free energy of the test liquid, and γ_S is the total surface free energy of CFRP. Through surface contact angle and surface energy analysis, the effects of low-temperature plasma treatment on the wettability and chemical activity of the CFRP surface were investigated.

3. Results and Discussion

3.1. *Results of Complex Experiment and Range Analysis*

Table 7 presents the shear strength of CFRP/Al7075 lap joints with different LTP surface treatment parameters (velocity, distance, and number of repeat scans) in complex experiments; the numbers 1, 2, and 3 in parentheses represent the different levels of each parameter. The results indicate that the shear strength of the joint in experiment T5 was the highest, reaching 26.44 MPa, which represents a 74.29% improvement compared to the untreated group. The LTP surface treatment parameters for this condition were a velocity of 10 mm/s, a distance of 15 mm, and three repeat scans.

Table 7. The results of the complex experiment.

No.	A	B	C	Average Shear Strength (MPa)	Standard Deviation
	Velocity (mm/s)	Distance (mm)	Number of Repeat Scans		
Untreated	/	/	/	15.17	0.32
T1	5 (1)	10 (1)	1 (1)	25.28	0.49
T2	5	15 (2)	2 (2)	23.63	0.23
T3	5	20 (3)	3 (3)	21.96	0.13
T4	10 (2)	10	2	23.71	0.42
T5	10	15	3	26.44	0.50
T6	10	20	1	22.61	0.50
T7	15 (3)	10	3	24.65	0.32
T8	15	15	1	24.86	0.44
T9	15	20	2	22.06	0.32

The range analysis of the complex experiment results is presented in Table 8. In the table, K_{mn} represents the sum of the shear strength of the joint at level n for factor m, while k_{mn} represents the arithmetic mean of K_{mn}. The range of factor m is denoted as R_m and is calculated using Equation (4). A larger value of R_m indicates a greater influence of factor m on the experimental results.

$$R_m = \max(k_{m1}, k_{m2}, \cdots, k_{mn}) - \min(k_{m1}, k_{m2}, \cdots, k_{mn}) \quad (4)$$

Table 8. The range analysis of the complex experiment.

	A	B	C
	Velocity (mm/s)	Distance (mm)	Number of Repeat Scans
K_1	70.88	73.64	72.76
K_2	72.76	74.93	69.40
K_3	71.57	66.64	73.05
k_1	23.63	24.55	24.25
k_2	24.25	24.98	23.13
k_3	23.86	22.21	24.35
R	0.63	2.77	1.21
Ranking		B > C > A	

The range analysis results indicate that the factors have the following order of impact on the shear strength of the joint: distance > number of repeat scans > velocity. Among these factors, distance (B) has the most significant impact on the shear strength of the joint.

3.2. Optimization of Main Influencing Parameters

Based on the range analysis results, it is evident that distance has the most significant impact on the shear strength of the joint. Therefore, the favorable experimental parameter T5 from the complex experiment is selected. By keeping the velocity and number of repeat scans constant and varying the distance, further optimization of the distance is conducted. The experimental plan is presented in Table 9.

Table 9. Optimization experiments of distance.

No.	A	B	C
	Velocity (mm/s)	Distance (mm)	Number of Repeat Scans
1	10	20	3
2	10	15	3
3	10	10	3
4	10	5	3

The shear strength of CFRP/Al7075 lap joints with different LTP surface treatment distances is shown in Figure 5.

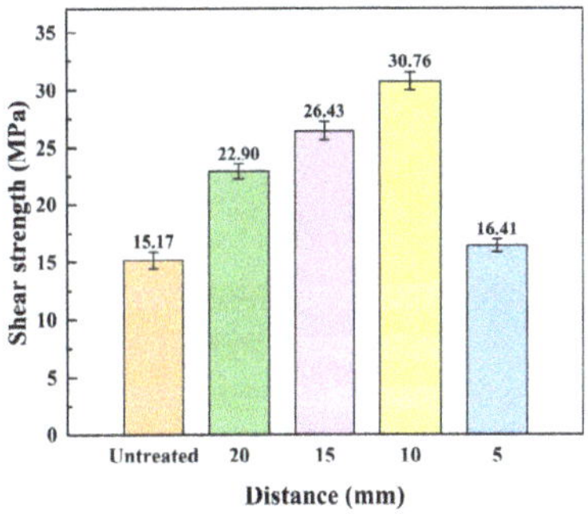

Figure 5. Shear strength of CFRP/Al7075 adhesive joints with different LTP surface treatment distances.

Figure 5 shows that after LTP surface treatment at a distance of 20 mm, the shear strength of CFRP/Al7075 lap joints significantly increased compared to the untreated group, with an improvement of approximately 50.96%. As the distance decreases, the shear strength of the lap joint exhibits a trend of initially increasing and then decreasing. At a distance of 10 mm, the shear strength of the lap joint reaches a peak value of 30.76 MPa. However, when the distance is further reduced to 5 mm, the shear strength of the joint drastically drops to 16.41 MPa. Based on this observation, it can be concluded that the CFRP/Al7075 lap joint exhibits the best bonding properties when the LTP surface treatment parameters are set as follows: a velocity of 10 mm/s, a distance of 10 mm, and three repeat scans. When the distance is further reduced, the LTP jet may cause excessive ablation of the resin on the surface of CFRP, damaging the properties of CFRP and resulting in a significant decrease in the shear strength of the joint.

The load-displacement curves of CFRP/Al7075 single-lap joints with different LTP surface treatment distances are shown in Figure 6.

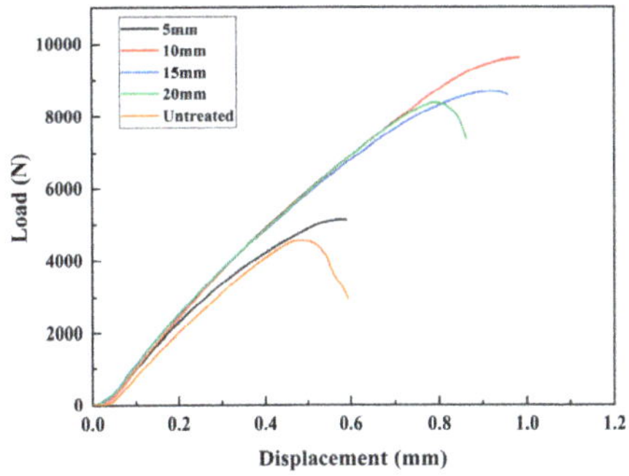

Figure 6. Load-displacement curves of CFRP/Al7075 single-lap joints with different LTP surface treatment distances.

Figure 6 shows that the CFRP/Al7075 single-lap joints treated with LTP surface treatment exhibit higher elastic moduli compared to the untreated group. This indicates that the interface bonding strength between the CFRP surface and the adhesive layer is enhanced after undergoing LTP surface treatment. When the distance is 10 mm, 15 mm, or 20 mm, the elastic moduli of the joints remain at a high level, and the peak load continuously increases with the increase in failure displacement. The joint with a distance of 5 mm and the untreated group both exhibit lower elastic moduli and lower peak loads. However, the reasons for this phenomenon differ between the two. The decrease in elastic modulus of the joint at a distance of 5 mm may be attributed to the excessive burning of the surface resin and fibers by the LTP process, leading to rapid debonding of the surface fibers and a significant reduction in joint shear strength. The lower elastic modulus of the untreated joint is due to the low interface bonding strength, resulting in a noticeable ductile failure process observed in the load–displacement curve.

3.3. Failure Mode Analysis

The typical failure modes of adhesive joints are primarily categorized into three types: substrate failure, interface failure, and adhesive failure. Under normal circumstances, adhesive joints can exhibit two or more failure modes, known as mixed failure. The different failure modes observed reflect the strength relationships between the substrate, adhesive, and interface [25].

Among the three failure modes, substrate failure is an ideal failure mode because it signifies that the adhesive bonding strength exceeds that of the substrate itself. This indicates that the adhesive bonding strength meets the required standards. Cohesive failure refers to separation within the adhesive layer itself, indicating that the adhesive's strength is lower than the interface bonding strength between the adhesive and the substrate. Interface failure refers to separation between the adhesive and the substrate at the bonding interface. If interface failure occurs in a joint, it typically indicates lower strength.

The failure modes of CFRP/Al7075 adhesive joints with different LTP surface treatment distances are shown in Figure 7.

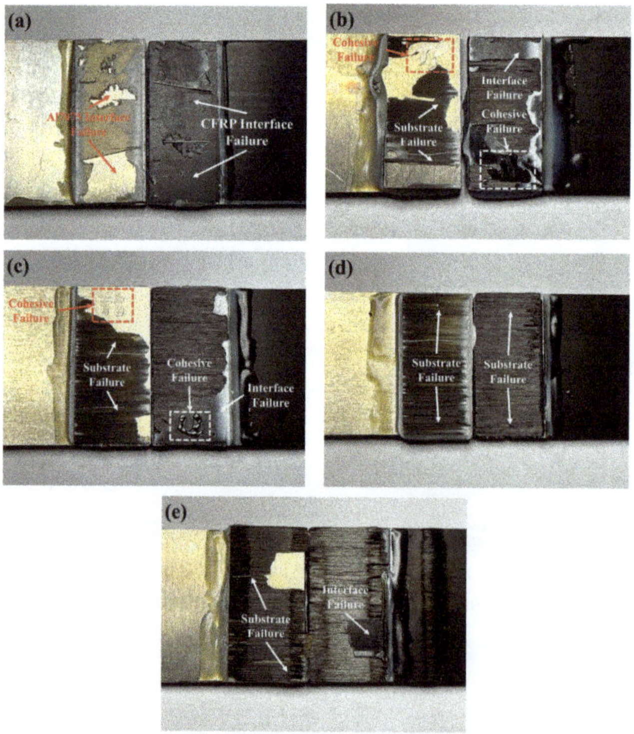

Figure 7. Failure modes of CFRP/Al7075 adhesive joints with different LTP surface treatment distances (**a**) Untreated (**b**) 20 mm (**c**) 15 mm (**d**) 10 mm (**e**) 5 mm.

Figure 7 shows that the failure mode of the untreated group joint is interface failure. In this case, the interface bonding strength between CFRP and the adhesive is weaker than that of Al7075, resulting in lower shear strength of the joint. At a distance of 20 mm, larger areas of CFRP substrate failure, as well as small areas of interface failure and cohesive failure, begin to appear on the fracture surface. This indicates that after LTP surface treatment, the interface bonding strength between CFRP and the adhesive is significantly improved. Consequently, the shear strength of the joint is substantially increased. At a distance of 15 mm, the area of CFRP substrate failure increases further, and the shear strength of the joint continues to increase as well. At a distance of 10 mm, the fracture surface shows complete CFRP substrate failure, and the shear strength of the joint reaches its peak. At a distance of 5 mm, there is also extensive CFRP substrate failure on the fracture surface. However, at this distance, the damaged CFRP surface exhibits severe burn marks, indicating that LTP has had a significant negative impact on the performance of CFRP itself. This also explains the sudden decrease in the shear strength of the joint.

In general, the changes in the failure mode of the joint correspond closely to the changes in the shear strength of the joints. LTP surface treatment primarily improves the bonding properties of the joint by enhancing the interface bonding strength between the CFRP surface and the adhesive.

3.4. Surface Wettability Analysis

Research has shown that the type of liquid, the microstructure, and the surface free energy of the material, as well as different interfaces, can all influence the wettability of materials. Composites with good wettability have a higher ability to be infiltrated by adhesives, resulting in the formation of high-strength bonding interfaces [26,27]. Therefore, in this study, the contact angles of CFRP surfaces with water and glycol were measured to characterize the surface wettability of CFRP.

The contact angles between the CFRP surface and water and glycol at different LTP surface treatment distances are shown in Figure 8.

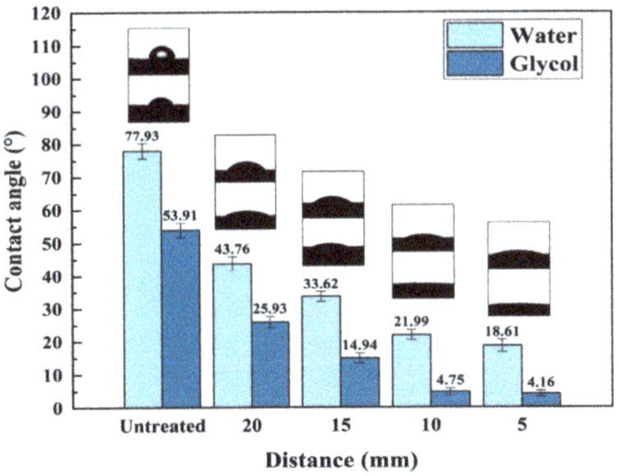

Figure 8. Contact angle between CFRP surface with water and glycol at different LTP surface treatment distances.

From Figure 8, it is evident that the contact angles between the CFRP surface of the bank group and the two test liquids are relatively large. However, after undergoing LTP surface treatment, there is a significant decrease in the contact angles of the CFRP surface. As the distance decreases, the reduction in contact angles becomes gradually smaller. When the distance reaches 10 mm and 5 mm, the contact angles remain nearly constant.

According to the Owens-Wendt-Rabel-Kaelble method [24], the dispersion component, polar component, and total surface free energy of the CFRP surface at different LTP treatment distances are calculated, as shown in Figure 9.

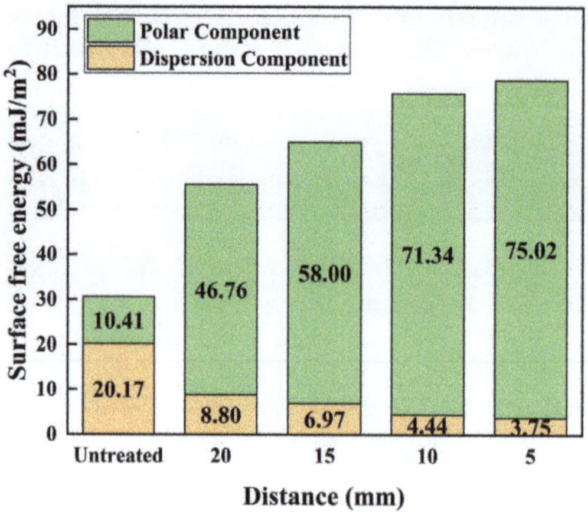

Figure 9. Dispersion component, polarity component, and total surface free energy of CFRP with different LTP surface treatment distances.

The surface free energy of polymers is primarily composed of polar and dispersion components. The polar component refers to the energy of interaction between polar molecules on the material surface, while the dispersion component refers to the energy of interaction caused by the instantaneous dipoles of the surface molecules (van der Waals forces). Changes in the polar and dispersion components reflect variations in the surface chemical activity of the material [28].From Figure 9, it can be observed that the dispersion component of the CFRP surface in the untreated group accounts for a higher proportion of the total surface energy, and the total surface energy is low. After LTP surface treatment, the polar component of the CFRP surface significantly increases, while the dispersive component slightly decreases. As a result, the total surface energy of CFRP experiences a significant increase. As the distance decreases, the proportion of the polar component in the surface free energy of CFRP increases, while the dispersive component continues to decrease. After reaching a distance of 10 mm, the magnitude of the increase in the polar component diminishes significantly, and the surface free energy no longer shows a noticeable improvement. This indicates that there is a saturation point for the improvement of surface wettability of CFRP by LTP surface treatment. Once the surface wettability of CFRP reaches this saturation point, further reduction in distance will have no significant impact on the surface energy of CFRP.

3.5. Surface Morphology Analysis

According to the theory of mechanical interlocking, when an adhesive flows over the surface of a material, it diffuses and infiltrates into small pores on the surface, displacing any air present. Once the adhesive cures, a mechanical interlock is formed between the adhesive and the material surface. Therefore, the surface morphology of the material also has a significant effect on the bonding strength [29]. The surface micromorphology of CFRP with different LTP surface treatment distances is shown in Figure 10.

Figure 10 shows that the surface of the CFRP in the untreated group appears smooth and flat, with good bonding between the fibers and the resin. There are also a few resin particles present on the surface. At a distance of 20 mm, the LTP surface treatment has caused a slight ablation on the CFRP surface, leading to the exposure of bare carbon fibers. Additionally, there is an increase in the presence of small granular resins. At a distance of 15 mm, the LTP surface treatment further deepens the ablation on the CFRP surface. In addition to the small granular resins, larger areas of pits are now appearing on the CFRP surface. At a distance of 10 mm, significant changes in the morphology of the CFRP surface are observed. Gully like textures are now distributed across the entire surface of the CFRP.

At a distance of 5 mm, the resin on the surface of the CFRP undergoes more severe ablation, leading to localized fiber debonding on the outermost layer.

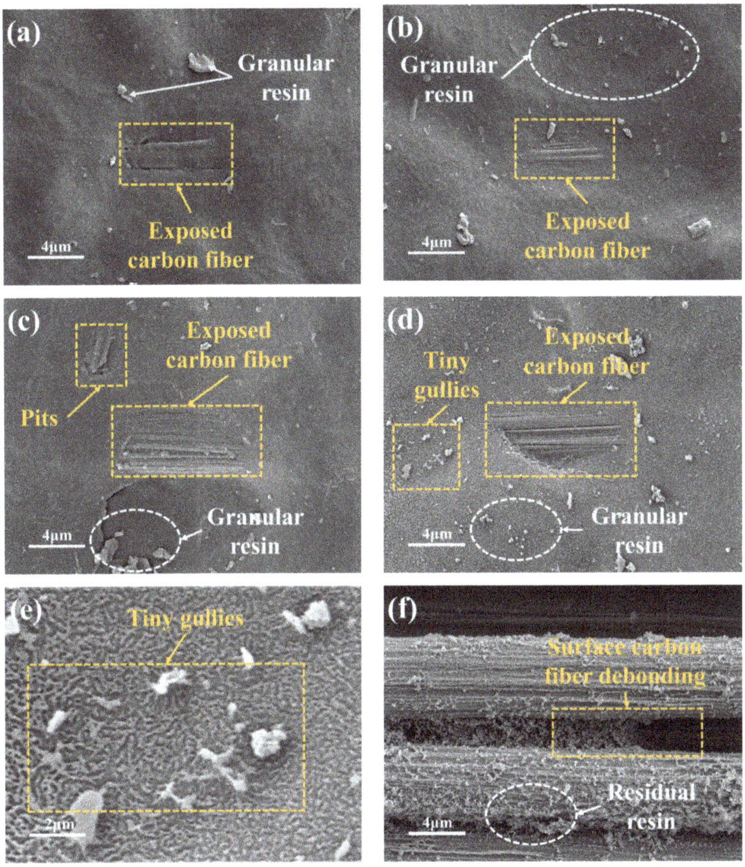

Figure 10. Surface micromorphology of CFRP with different LTP surface treatment distances (**a**) Untreated (**b**) 20 mm (**c**) 15 mm (**d**) 10 mm (**e**) Partial Enlarged Image of 10 mm group (**f**) 5 mm.

In general, when the LTP surface treatment distance is 20 mm or 15 mm, the ablation depth of the CFRP surface is not significant. The surface morphology remains relatively flat, showing minimal difference compared to the CFRP from the untreated group. At a distance of 10 mm, a significant change in the morphology of the CFRP surface occurs, resulting in gully like textures. These textures contribute to enhancing the mechanical interlocking between the adhesive and the CFRP surface, thereby resulting in the highest shear strength of the joint. At a distance of 5 mm, the LTP surface treatment reaches a critical point for the ablation of CFRP. At this point, the resin is severely ablated, and localized fiber debonding occurs, resulting in a significant decrease in the shear strength of the joint.

3.6. Surface Chemical Structure Analysis

Research has shown that the excited-state radicals present in LTP carry a significant amount of energy, capable of breaking any organic chemical bond. Through LTP surface treatment, a series of complex grafting and crosslinking reactions occur on the CFRP surface, leading to changes in the surface's chemical structure. The fracture of certain organic bonds in CFRP generates new reactive groups, thereby enhancing the surface's chemical activity [17]. The relative elemental content and XPS wide-scan spectrum of CFRP surfaces with different LTP surface treatment distances are shown in Table 10 and Figure 11.

Table 10. Relative elemental content of CFRP with different LTP surface treatment distances.

Distance	Relative Content of Elements (Atomic %)					
	C	O	N	Si	O/C	N/C
Untreated	90.72	4.55	0.87	3.86	0.05	0.01
20 mm	64.91	27.03	5.32	1.67	0.42	0.08
15 mm	64.25	28.6	4.36	2.28	0.45	0.07
10 mm	58.55	33.6	3.33	2.53	0.57	0.06
5 mm	55.71	37.37	4.76	2.16	0.67	0.08

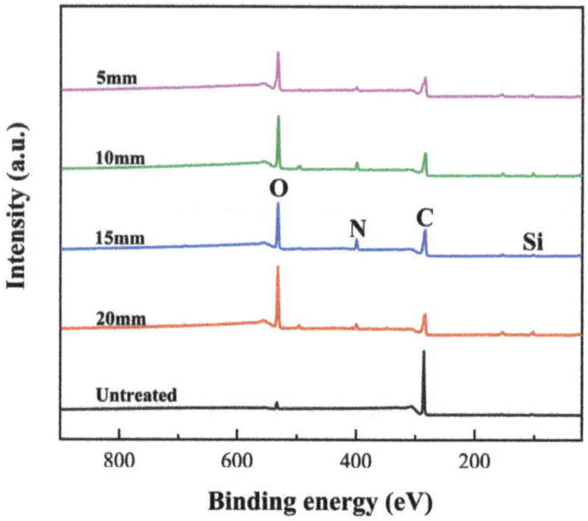

Figure 11. XPS wide-scan spectrum of CFRP with different LTP surface treatment distances.

Based on Figure 11, it can be seen that the surface of the control group's CFRP primarily consists of carbon, and the content of nitrogen and oxygen is relatively low. After LTP surface treatment at a distance of 20 mm, the oxygen content on the surface of CFRP significantly increases, while the nitrogen content also shows a slight increase. As the distance decreases, the oxygen-to-carbon ratio shows a continuous increase, while the nitrogen-to-carbon ratio does not show significant changes. This indicates that changes in the LTP surface treatment distance primarily affect the content of oxygen-containing functional groups on the surface of CFRP, while having a minimal impact on nitrogen-containing functional groups. At the same time, the silicon content on the surface of CFRP with different LTP surface treatment distances exhibits an irregular variation trend, which may be related to the use of silicon-containing release agents during the CFRP fabrication process.

The results of C1s peak fitting of CFRP with different LTP surface treatment distances are shown in Figure 12. The black scattered points in the figure represent the raw data, the red curve represents the fitted curve, and the gray curve represents the baseline. The absorption peaks of different functional groups are represented by curves of different colors.

The results of the C1s peak fitting indicate that the oxygen-containing radicals in LTP primarily react with the C=C and C-C groups on the surface of CFRP, resulting in the formation of oxygen-containing functional groups such as C-O, C=O, and C-C=O. After the LTP surface treatment, the C=C groups on the surface of CFRP in the untreated group disappear and are converted into oxygen-containing functional groups. Furthermore, as the distance decreases, the C-C groups also continuously react with oxygen-containing radicals, leading to the generation of oxygen-containing functional groups. This continuous reaction process increases the oxygen-to-carbon ratio on the surface of CFRP.

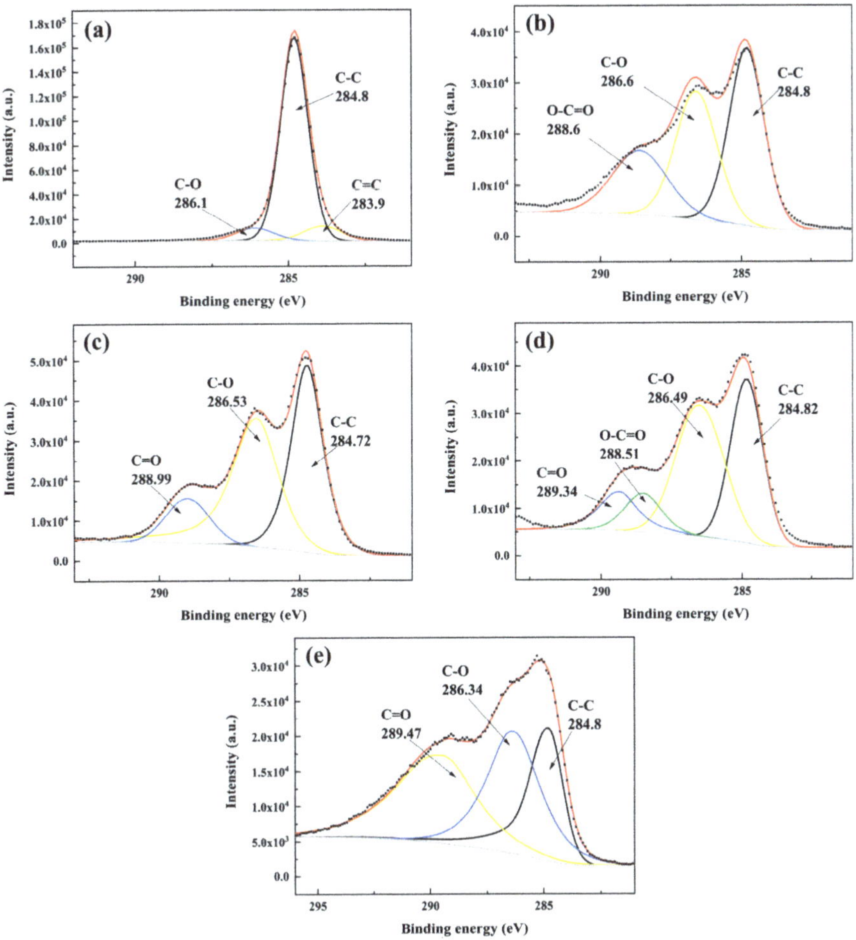

Figure 12. Results of C1s peak fitting of CFRP with different LTP surface treatment distances (**a**) Untreated (**b**) 20 mm (**c**) 15 mm (**d**) 10 mm (**e**) 5 mm.

The relative content of C1s groups on the surface of CFRP with different LTP surface treatment distances is shown in Table 11.

Table 11. The relative content of C1s groups on the surface of CFRP with different LTP surface treatment distances.

Distance	Relative Content of C1s Groups (Atomic %)					Polar Groups/Non-Polar Groups
	C=C	C-C	C-O	C=O	O-C=O	
Untreated	8.70	83.97	7.33	/	/	0.07
20 mm	/	41.62	34.78	/	23.60	1.40
15 mm	/	42.02	45.39	12.59	/	1.41
10 mm	/	36.05	42.22	11.70	10.03	1.77
5 mm	/	25.95	35.78	38.46	/	2.85

According to Table 11, it is evident that the types and quantities of C1s functional groups on the surface of CFRP undergo changes with different LTP surface treatment distances. The LTP surface treatment primarily increases the content of oxygen-containing functional groups on the surface of CFRP. As the treatment distance decreases, the oxygen-containing radicals continuously react with the C-C groups, leading to the formation of oxygen-containing functional groups. The proportion of polar functional groups to non-polar functional groups on the surface of CFRP increases from an initial value of 0.07 to 2.85. This significant increase in the ratio indicates a substantial improvement in surface chemical activity.

Overall, LTP surface treatment significantly improves the surface chemical activity of CFRP. As the distance decreases, the proportion of oxygen-containing polar functional groups on the surface of CFRP continuously increases. This enhancement effectively improves the surface energy and wettability of CFRP, ultimately leading to improved interface bonding properties with adhesives.

4. Conclusions

(1) The results of the complex experiment and range analysis indicate that the ranking of the impact of the three parameters on the shear strength of the joint is as follows: distance > number of repeat scans > velocity. At a velocity of 10 mm/s, a distance of 10 mm, and three repeat scans, the joint shows the highest shear strength of 30.76 MPa, which is a 102.76% improvement compared to the joints in the untreated group. Interface failure between CFRP and the adhesive is the main cause of low strength in the joints of the untreated group. The LTP surface treatment effectively improves the interface bonding strength between CFRP and the adhesive;

(2) LTP surface treatment significantly improves the wettability of CFRP compared to the untreated group. However, as the treatment distance decreases, there is a saturation value for the surface wettability of CFRP. Once the saturation value is reached, the wettability of CFRP no longer significantly improves;

(3) When the treatment distance is 10 mm, the CFRP surface exhibits gully like textures, which enhance the interface bonding strength between CFRP and the adhesive through mechanical interlocking. When the treatment distance is 5 mm, the surface resin of the CFRP experiences severe ablation, leading to localized fiber debonding;

(4) After LTP surface treatment, the oxygen content on the surface of CFRP significantly increases. As the treatment distance decreases, the proportion of oxygen-containing functional groups increases, leading to higher chemical reactivity of the CFRP surface.

Author Contributions: Writing—review and editing, L.W.; methodology, resources, writing—review and editing, R.W.; writing—original draft, E.X. All authors have read and agreed to the published version of the manuscript.

Funding: This research received no external funding.

Institutional Review Board Statement: Not applicable.

Informed Consent Statement: Informed consent was obtained from all subjects involved in the study.

Data Availability Statement: Data are contained within the article.

Conflicts of Interest: The authors declare no conflicts of interest.

References

1. Das, T.K.; Ghosh, P.; Das, N.C. Preparation, Development, Outcomes, and Application Versatility of Carbon Fiber-Based Polym Composites: A Review. *Adv. Compos. Hybrid. Mater.* **2019**, *2*, 214–233. [CrossRef]
2. McIlhagger, A.; Archer, E.; McIlhagger, R. Manufacturing Processes for Composite Materials and Components for Aerospa Applications. In *Polymer Composites in the Aerospace Industry*; Elsevier: Amsterdam, The Netherlands, 2020; pp. 59–81. [CrossR
3. André, N.M.; Bouali, A.; Maawad, E.; Staron, P.; dos Santos, J.F.; Zheludkevich, M.L.; Amancio-Filho, S.T. Corrosion Behavior Metal–Composite Hybrid Joints: Influence of Precipitation State and Bonding Zones. *Corros. Sci.* **2019**, *158*, 108075. [CrossRe

Hosseinabadi, O.F.; Khedmati, M.R. A Review on Ultimate Strength of Aluminium Structural Elements and Systems for Marine Applications. *Ocean. Eng.* **2021**, *232*, 109153. [CrossRef]
Miguel Silva, J.; Ferreira, F.; Abreu, S.M.; Matos, J.E.; Durão, L.M.P. Correlation of Drilling Damage with Mechanical Strength: A Geometrical Approach. *Compos. Struct.* **2017**, *181*, 306–314. [CrossRef]
Ranjan, P.; Pandey, A.K. Modeling of Pinning Phenomenon in Iwan Model for Bolted Joint. *Tribol. Int.* **2021**, *161*, 107071. [CrossRef]
Jiang, H.; Sun, L.; Dong, D.; Li, G.; Cui, J. Microstructure and Mechanical Property Evolution of CFRP/Al Electromagnetic Riveted Lap Joint in a Severe Condition. *Eng. Struct.* **2019**, *180*, 181–191. [CrossRef]
Rahman, M.; Rabby, M.M.; Vadlamudi, V.; Raihan, R. A Multi-Parameter Approach to Predicting Adhesive Bond Strength in Carbon Fiber Reinforced Composites through Surface and Dielectric Characterization. *Int. J. Adhes. Adhes.* **2024**, *128*, 103531. [CrossRef]
Jang, J.; Yang, H. Effect of Surface Treatment on the Performance Improvement of Carbon Fiber/Polybenzoxazine Composites. *J. Mater. Sci.* **2000**, *35*, 2297–2303. [CrossRef]
Ledesma, R.I.; Palmieri, F.L.; Lin, Y.; Belcher, M.A.; Ferriell, D.R.; Thomas, S.K.; Connell, J.W. Picosecond Laser Surface Treatment and Analysis of Thermoplastic Composites for Structural Adhesive Bonding. *Compos. B Eng.* **2020**, *191*, 107939. [CrossRef]
Ehrmann, C.; Li, N.; Wen, L.; Xu, X.; Qin, L. Effect of Low-Temperature Plasma Surface Treatment on Bonding Properties of Single-Lap Joint of Thermosetting Composites. *Polymers* **2023**, *15*, 1631. [CrossRef]
Encinas, N.; Díaz-Benito, B.; Abenojar, J.; Martínez, M.A. Extreme Durability of Wettability Changes on Polyolefin Surfaces by Atmospheric Pressure Plasma Torch. *Surf. Coat. Technol.* **2010**, *205*, 396–402. [CrossRef]
Ataeefard, M.; Moradian, S.; Mirabedini, M.; Ebrahimi, M.; Asiaban, S. Investigating the Effect of Power/Time in the Wettability of Ar and O2 Gas Plasma-Treated Low-Density Polyethylene. *Prog. Org. Coat.* **2009**, *64*, 482–488. [CrossRef]
Lennon, P.; Espuche, E.; Sautereau, H.; Sage, D. Influence of Microwave Plasma Treatment on the Wettability and the Adhesive Properties of Polyamides Films with an Epoxy Resin. *Int. J. Adhes. Adhes.* **1999**, *19*, 273–279. [CrossRef]
Baba, E.M.; Cansoy, C.E.; Zayim, E.O. Optical and Wettability Properties of Polymers with Varying Surface Energies. *Appl. Surf. Sci.* **2015**, *350*, 115–120. [CrossRef]
Redmer, R. Physical Properties of Dense, Low-Temperature Plasmas. *Phys. Rep.* **1997**, *282*, 35–157. [CrossRef]
Foest, R.; Schmidt, M.; Becker, K. Microplasmas, an Emerging Field of Low-Temperature Plasma Science and Technology. *Int. J. Mass. Spectrom.* **2006**, *248*, 87–102. [CrossRef]
Rahman, T.U.; Roy, H.; Fariha, A.; Shoronika, A.Z.; Al-Mamun, M.R.; Islam, S.Z.; Islam, M.S.; Marwani, H.M.; Islam, A.; Alsukaibi, A.K.D.; et al. Progress in Plasma Doping Semiconductor Photocatalysts for Efficient Pollutant Remediation and Hydrogen Generation. *Sep. Purif. Technol.* **2023**, *320*, 124141. [CrossRef]
Liu, S.H.; Trelles, J.P.; Li, C.J.; Li, C.X.; Guo, H.B. A Review and Progress of Multiphase Flows in Atmospheric and Low Pressure Plasma Spray Advanced Coating. *Mater. Today Phys.* **2022**, *27*, 100832. [CrossRef]
Li, C.; Viswanathan-Chettiar, S.; Sun, F.; Shi, Z.; Blackman, B. Effect of CFRP Surface Topography on the Adhesion and Strength of Composite-Composite and Composite-Metal Joints. *Compos. Part. A Appl. Sci. Manuf.* **2023**, *164*, 107275. [CrossRef]
Kehrer, M.; Rottensteiner, A.; Hartl, W.; Duchoslav, J.; Thomas, S.; Stifter, D. Cold Atmospheric Pressure Plasma Treatment for Adhesion Improvement on Polypropylene Surfaces. *Surf. Coat. Technol.* **2020**, *403*, 126389. [CrossRef]
Rafailović, L.D.; Stupavska, M.; Prysiazhnyi, V.; Polt, G.; Rohr, O.; Nixon, S.; Rahel, J. Impact of Atmospheric Pressure Plasma Treatment on Surface Metallization of CFRP Composites. *Surf. Coat. Technol.* **2021**, *412*, 127046. [CrossRef]
Dighton, C.; Rezai, A.; Ogin, S.L.; Watts, J.F. Atmospheric Plasma Treatment of CFRP Composites to Enhance Structural Bonding Investigated Using Surface Analytical Techniques. *Int. J. Adhes. Adhes.* **2019**, *91*, 142–149. [CrossRef]
Waldner, C.; Hirn, U. Modeling Liquid Penetration into Porous Materials Based on Substrate and Liquid Surface Energies. *J. Colloid. Interface Sci.* **2023**, *640*, 445–455. [CrossRef] [PubMed]
Pan, W.; Sun, L.; Wei, J.; Zhang, Y.; Gao, X.; Wang, J. Fatigue Failure Prediction Method and Constant Amplitude Equivalence Strategy of Adhesive Interface under Variable Amplitude Loading. *Int. J. Adhes. Adhes.* **2024**, *129*, 103586. [CrossRef]
Bozaci, E.; Sever, K.; Sarikanat, M.; Seki, Y.; Demir, A.; Ozdogan, E.; Tavman, I. Effects of the Atmospheric Plasma Treatments on Surface and Mechanical Properties of Flax Fiber and Adhesion between Fiber-Matrix for Composite Materials. *Compos. B Eng.* **2013**, *45*, 565–572. [CrossRef]
Zhu, Z.; Zheng, S.; Peng, S.; Zhao, Y.; Tian, Y. Superlyophilic Interfaces and Their Applications. *Adv. Mater.* **2017**, *29*, 1703120. [CrossRef] [PubMed]
Yamauchi, K.; Shiotani, Y.; Saito, H. Deviation from Linear Relationship between Adhesion Strength and Surface Free Energy on a Low Energy Surface. *Procedia Struct. Integr.* **2023**, *45*, 125–131. [CrossRef]
Hamilton, A.; Xu, Y.; Kartal, M.E.; Gadegaard, N.; Mulvihill, D.M. Enhancing Strength and Toughness of Adhesive Joints via Micro-Structured Mechanical Interlocking. *Int. J. Adhes. Adhes.* **2021**, *105*, 102775. [CrossRef]

coatings

Article

The Effect of Resin Interleafing on the Wedge Peel Strength of CF/PEEK Manufactured by Laser-Assisted In Situ Consolidatio

Ruozhou Wang, Entao Xu and Liwei Wen *

College of Materials Science and Technology, Nanjing University of Aeronautics and Astronautics, Nanjing 210016, China
* Correspondence: wenliwei@nuaa.edu.cn

Abstract: In this work, a novel approach involving coating fine PEEK powder on prepreg is introduced to improve wedge peel strength and reduce interlaminar voids. CF/PEEK laminates with resin interleaving are in situ consolidated by laser-assisted fiber placement. The morphology of the powdered surface is obtained using an optical profilometer, and the surface roughness and volume of added resin are calculated accordingly. Interface and surface temperature are measured during the layup process. Thermal history indicates that very short bonding time is the dominating factor for voids and limited interlayer strength. Laminate porosity and microscopic features are characterized with an optical microscope. The porosity of resin-interleaved laminates decreases to 3.7%, while the resin content only increases by 4.5% in the meantime. This is because interlayer resin particles rapidly melt under laser heating and quickly fill the voids between layers. The wedge peel strength of resin-interleaved laminates can increase by 30.1% without a repass treatment. This could be attributed to the increase in resin intimate contact and reduction in interlayer voids.

Keywords: thermoplastic composite; laser-assisted forming; automated fiber placement (AFP); microstructures; mechanical properties

1. Introduction

The application of thermoplastic composites is gradually expanding. Their advantageous properties, including lightweight, short cycle time, and recyclability, have attracted attention in the aerospace industry [1–3]. Laser-assisted fiber placement (LAFP) is a promising additive technique that the thermoplastic prepreg can be consolidated in situ during the layup process. However, the performance of laminates manufactured by AFP currently lags behind those produced by traditional methods like molding and autoclaving. Minimizing this performance gap is the focus of recent studies [4,5].

The impact of processing parameters in LAFP has been thoroughly investigated, including laser power, nip-point temperature, compaction force, laying speed, mold temperature, etc. [6–10]. However, the lower strength is essentially attributed to interlayer voids, because the consolidation time between layers is usually less than 1 s during layup. This characteristic of LAFP cannot be altered.

According to the fusion bonding model proposed by Lee W. et al. [11,12], a low degree of resin contact leads to interlaminar voids and consequently lower bonding strength. The exposed dry fibers on the thermoplastic tape surface are obstacles between layers [13–15]. A direct manner to reduce such voids is by adding resin between layers. In the work of Qureshi [4], Hulcher [16], and Miao [17], resin film interleaved thermoplastic laminates manufactured by AFP have shown better properties in peel strength and interlaminar shear strength (ILSS), respectively. However, further thermal treatment like repass treatment is needed to melt the added resin film. Repass treatment means reheating and pressing the current layer by the AFP machine without laying new tape [18,19]. This process comes at the cost of production efficiency. Moreover, the resin film used usually has a thickness of 15–100 μm [17,20]. Fiber-dominated properties are weakened because of the increase in resin volume content [21].

In this work, a novel approach involving coating fine PEEK powder on prepreg is introduced to enhance wedge peel strength and reduce interlaminar voids. Unlike resin films, the added resin powder melts more rapidly during laser heating, eliminating the necessity for subsequent heat treatment. Moreover, minimal amounts of resin powder can be added, ensuring the mechanical properties of the laminate. This study investigates the surface morphology of prepreg with coated PEEK powder and explores the thermal history during layup. The feasibility of adding PEEK powder to improve the wedge peel strength is discussed.

2. Materials and Methods

2.1. Material and Equipment

A unidirectional thermoplastic prepreg Cetex TC1200 AS-4D/PEEK (TenCate Co., Ltd., Nijverdal, The Netherland)was used in this work. CF/PEEK has excellent mechanical and heat resistance properties. It has been commonly used for LAFP-related research work [6,8–10]. The tape is 12 inches in width when received and is slit to 6.35 mm for fiber placement. The relevant properties of the as-received tape are listed in Table 1.

Table 1. Properties of the as-received AS4D/PEEK tape.

rea Density/g/m^2	Resin Mass Content/%	Thickness/mm	Glass Transformation Temperature/°C	Melting Temperature/°C	Void Content/%
223	34	0.143	143	343	0.7

Placement trials were performed using a laser ATP system designed by the authors, as depicted in Figure 1.

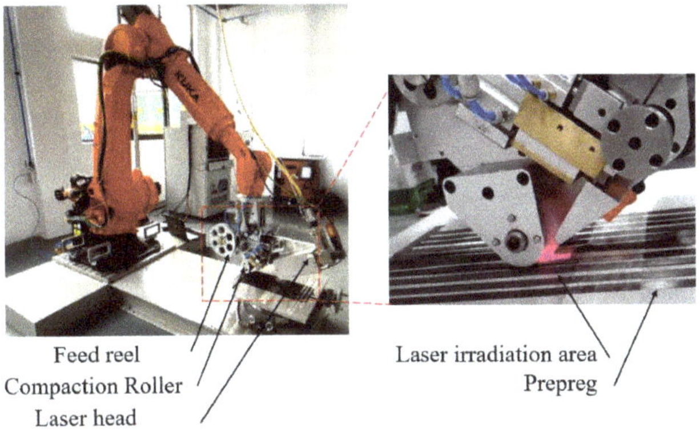

Figure 1. Laser AFP equipment used in this study.

The laser ATP head was mounted on an industrial robot (KUKA KP2700). A 2 kW near-infrared laser source (Raycus Fiber Laser Technology Co., Wuhan, China) ensured the energy needed for the layup process. The laser power was sufficient to heat the material to its melting temperature (343 °C). A customized integrating lens caused the laser beam to focus on the nip point with spot sizes of 7 × 20 mm. A coaxial temperature sensor (Sensortherm sensor, Steinbach, Germany) was integrated into the lens to detect laser spot temperature. Due to the high energy density of the laser, the material heating rate was rapid, so the laser power was close-loop-regulated by a control system (Sensortherm M serious controller) according to its rapid PID algorithm. In this way, the nip-point temperature remained constant during the layup process.

A Fluoro rubber roller with 55 shore hardness (Yangzhong Rubber and Plastics Co., Yangzhong, China) was used to provide sufficient consolidation force at elevated temperatures. A torque motor attached to the feeding reel could adjust the tow tension to the set value. The prepreg surface temperature and interlaminar temperature during the layup process were measured with an infrared camera (FOTRIC Infrared Technology Co., Dallas, TX, USA) and a K-type thermocouple (0.1 mm in diameter), respectively.

The key parameters of the LAFP equipment are listed in Table 2.

Table 2. Key parameters of the LAFP equipment.

Part Name	Part Parameter
Roller diameter	34 mm
Compaction force	10–150 N
Layup speed	10–200 mm/s
Laser wavelength	1064 nm
Laser irradiation area	7 × 20 mm
Lens focal length	400 mm

2.2. Resin Powder Interleafing

Resin film (15–100 μm) could be added between layers to enhance interlaminar strength. However, resin-rich interleaves reduce fiber-related properties such as tensile strength, as the fiber fraction is relatively decreased. In preliminary experiments, it was observed that the resin film applied to the surface of the prepreg was not easily melted during laser heating. This was attributed to the significantly higher absorption rate of carbon fibers compared to the resin, resulting in the rapid heating of the fibers and the slower heating of the resin film. Consequently, the film hindered interlaminar bonding rather than facilitating this process. Therefore, additional thermal treatment like repass treatment is necessary for film interleafing. However, such additional thermal treatment sacrifices laydown efficiency. Using powder coating helps reduce the amount of resin added, contributing to the mechanical properties of the component. Moreover, resin particles have a large surface area, making them more prone to melting under laser irradiation without any further treatment.

Victrex 90G PEEK powder was selected due to its low viscosity. This type of resin is produced by Victrex Company (Lancashire, UK) in the UK and is commonly used in injection molding. It exhibits the ability to rapidly fill interlayer voids once melted [22,23]. It is a high-performance thermoplastic polymer, semi-crystalline, depth-filtered granules for injection molding and other processes, with very easy flow properties. Typical application areas include complex geometries with thinner cross-sections or long flow lengths. It is also suitable for aggressive environments and food contact applications. The relevant properties of the PEEK powder are listed in Table 3. The average diameter of the powder was about 20 μm when received. The small size of the particles facilitated a large specific surface area, enabling rapid melting under laser heating.

Table 3. Relevant properties of the PEEK powder.

Material Data	Conditions	Typical Value
Melting temperature/°C		343
Glass transition (Tg)/°C		143
Thermal conductivity/$Wm^{-1}K^{-1}$	At 23 °C	0.29
Density g/cm^3	Crystalline	1.30
Density g/cm^3	Amorphous	1.27
Melt viscosity/pa·s	at 400 °C	90
Tensile strength/MPa		110
Tensile Modulus/GPa		3.9

The PEEK powder was coated onto the prepreg surface using an electrostatic spraying device. As shown in Figure 2, due to electrostatic attraction, the powder can stably adhere to the film surface. The amount of added powder was controlled by the air pressure during spraying. The weight of added power is calculated according to the mass difference of the prepreg before and after powder coating. Three different coating levels, namely low, medium, and high coating amounts, were set as experimental variables, designated as P1, P2, and P3, respectively. The control group in the experiment, which consisted of prepreg without resin addition, is labeled as NP (no powder). The precise quantities of powder addition are displayed in Table 4.

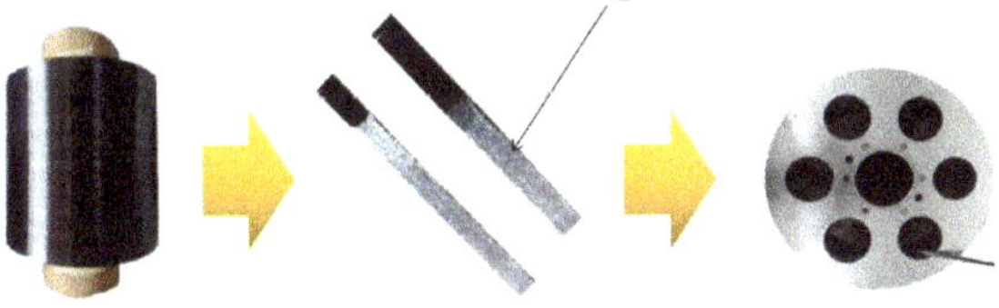

Figure 2. Schematic diagram of prepreg slit and powder spray.

Table 4. The amount of resin powder coated on prepreg.

Samples	Powder Layer Thickness in Average (μm)	Powder Weight (g/m^2)	Resin Mass Fraction (%)	Surface Area Covered by Resin (%)
NP	0	0	34	12.7
P1	6.1	7.9	38.5	25.5
P2	9.3	12.2	40.4	50.6
P3	16.6	21.6	44.7	81.2

2.3. *Placement Trials*

According to studies in the literature, higher temperatures, increased pressure, and appropriate fiber tension can yield optimal bonding strength. Based on the author's preliminary experiments, the optimal layup process parameters were a layup speed of 100 mm/s, a fixed nip temperature of 400 degrees, a layup pressure of 100 N, and a fiber tension of 8 N. These process parameters were compatible with the layup equipment in this paper and closely align with the recommendations in the literature [6,9,10,24].

A schematic diagram of laser-assisted forming with PEEK powder coating is shown in Figure 3a. A thermal insulation board was placed on the mold to prevent heat loss and promote interlayer bonding. The first prepreg layer was double-sided-taped onto a thermal insulation board, followed by subsequent layers being sequentially laid up to form a $[0]_4$ laminate. To facilitate peel testing, a small piece of aluminum foil was placed between the second and third layers as a pre-cut crack. Ultimately, an interleaved laminate with a size of 6.35 × 300 mm was produced, as depicted in Figure 3b.

Multiple specimens were prepared for each set of experimental parameters. Among these, five were used for peel testing, while one was used for cross-sectional profiles. After trimming the edges of the specimens, their widths were measured using a vernier caliper, with the specimen width maintained at 6.35 mm.

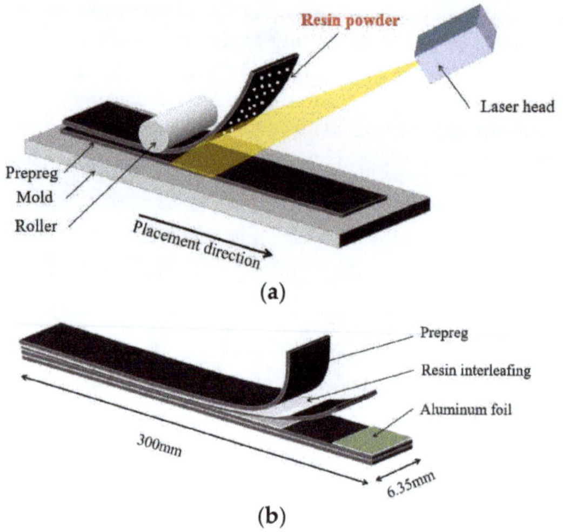

Figure 3. Schematic diagram of laser-assisted forming with PEEK powder coating (**a**) and manufactured samples (**b**).

2.4. Interlaminar Strength Characterization

The quality of the interlaminar bond is essential to product quality. Composite interlaminar properties are usually evaluated by short beam strength (SBS, ASTM D2344 [25]) and double-cantilever beam (DCB, ASTM D5528 [26]) tests. Both are sensitive to resin properties and are often utilized for benchmarking. However, plastic deformation in thermoplastic laminates during SBS testing may lead to complex failure modes and inaccurate interlaminar strength [6]. This is thought to be due to the voids and poor fiber–resin bonding in laminates.

The wedge peel test has been employed to evaluate the interlaminar properties of samples made by the AFP process [24,27]. The wedge peel test requires a specimen only 4–6 layers thick, while the DCB test requires the thickness to be between 3 and 5 mm, which means 24 to 40 layers, so the specimen preparation and testing of the peel test is more time-saving. The wedge peel test shares similar failure modes with the DCB test and shows a high correlation with test results. Peaks in both the wedge peel strength and DCB fracture toughness data are observed to occur well [28]. These favorable factors make it a commonly used quality indicator in AFP studies.

The wedge peel test equipment used in this study is illustrated in Figure 4. It has a clamp attached to a universal testing machine and a peeling blade separating the specimen. The peeling blade has a thickness of 3.2 mm and an angle of 30 degrees. The peel force was measured using a 200 N force sensor, while peel displacement was obtained through position sensors in the universal testing machine, both sampled at a frequency of 50 Hz. During the peel test, significant variability in peel force was observed at the initial and final peel ends of the specimen, and these data were discarded. Data from the relatively stable peel region in the middle of the specimen were selected for statistical analysis.

A SENSOFAR-SNEOX 3D optical profiler (Sensofar, Barcelona, Spain)was used to evaluate surface morphology so that the as-received tape surface and the coated surface morphology were compared. Furthermore, according to surface profile data, the average thickness of interleaved powder resin was calculated. The precise quantities of powder addition are displayed in Table 4. An integrated SNEOX 3D optical microscope (Sensofar, Barcelona, Spain)was used for the micrographs of the material surface and for observing the resin particle size.

Samples made by laser-assisted consolidation were ground and polished, and the polished surfaces of cross-sections were observed with an optical microscope to analyze the characteristics of voids and fiber–resin distribution. The optical microscope used for cross-section observation was Leica DVM5000 (Heidelberg, Germany). Porosity was calculated according to micrographs, with the help of ImageJ software v1.53.

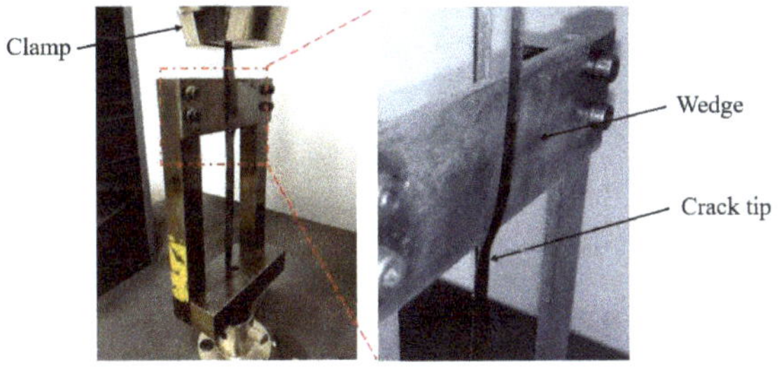

Figure 4. Image of wedge peel test device.2.5. Other Characterization Tests.

PEEK resin within the raw material undergoes fusion with the added PEEK powder during laser consolidation. To discern the interface between these constituents, a polarizing filter with a yellow color was attached to the optical microscope. Due to the optical differences between the two resins, the interface between them can be distinguished. This allows for the study of the flow behavior of PEEK powder during the forming process.

3. Results and Discussion

3.1. Surface Micrographs

For the experiment, different resin coating amounts were set as experimental variables, labeled P1, P2, P3, and NP (no resin added). Their 3D surface morphology images are shown in Figure 5. Figure 5a depicts the surface without coating, and it can be seen that the surface consists of many exposed fibers, while resin-covered areas are only 12.7%. Although the raw material contains 34% resin, these resins are concentrated within the material rather than on its surface. Therefore, the exposed fibers on the surface become obstacles to the molding process. It is these areas of exposed fibers that cause interlayer voids and lower interlayer strength. Figure 5b–d correspond to different levels of PEEK powder coating, with resin particle sizes approximately 20 μm. It can be observed that with the increase in PEEK powder coating amount, the size of exposed fibers decreases continuously, and the average thickness of resin on the prepreg surface increases.

Based on the surface morphology data, the average resin thickness and the proportion of resin-covered area on the surface of the prepreg under different coating amounts were calculated, as demonstrated in Table 4. This suggests that as the PEEK powder content increases, the proportion of exposed fibers on the prepreg surface decreases. Powder coating has a relatively minor impact on laminate resin fraction. The average thickness of the resin interleaf of samples P1 and P2 are only 6.1 and 9.3 μm, significantly thinner than the resin films reported in the literature. According to the measurement, the average resin thickness of P3 samples is 16.6 μm, which is close to the resin film thickness reported in the literature. This powdered interleaf has a relatively minor impact on the fiber content of the laminate while maintaining fiber-related mechanical properties, such as tensile strength.

According to the fusion bonding model, close contact between interlayer resins is essential for interlayer bonding. However, this process is impeded by exposed fibers in the LAFP process. In regions with abundant fibers on the prepreg surface, the resin must be squeezed into the interlayer from the interior of the prepreg. Achieving this squeezed resin flow is challenging for LAFP due to the resistance of fiber and the very limited process time. Powder coating dramatically increases the resin content on the surface of the prepreg, rising from 12.7% (as-received prepreg) to 81.2% in the P3 sample. Therefore, adding interlayer powder helps improve the intimate contact of the prepreg and enhances the interlayer bonding strength.

(a)

(b)

(c)

(d)

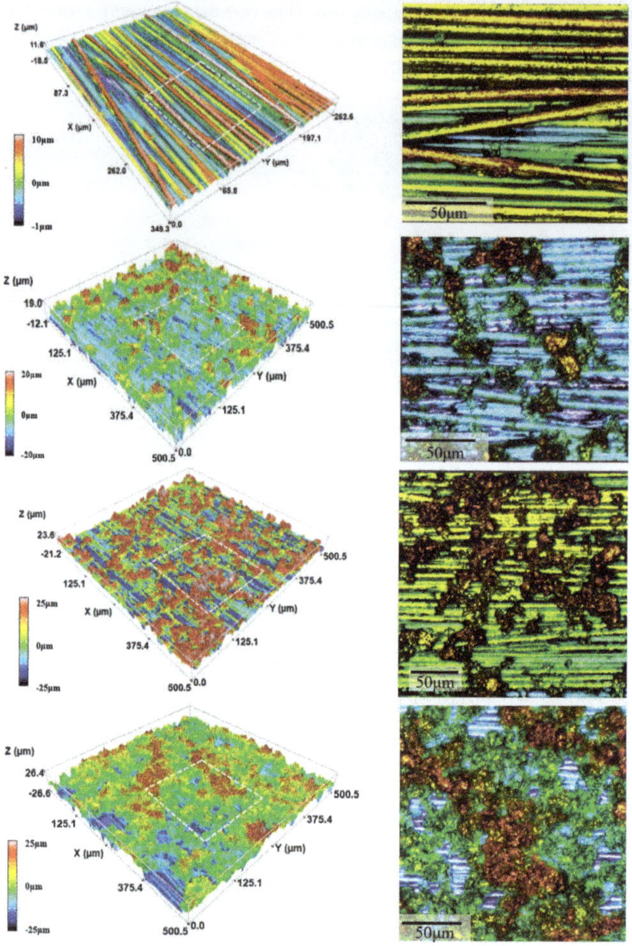

Figure 5. The 3D surface morphology images of the sample: (**a**) NP, (**b**) P1, (**c**) P2, and (**d**) P3.

3.2. Thermal History

In order to analyze the thermal history during the forming process, an infrared thermal imager and K-type thermocouples were used to measure the surface temperature of the prepreg and the interlayer temperature, respectively. The measurement results indicate that interlayer powder coating has no significant impact on the thermal history during processing. This thermal history test was conducted using a 1080 nm laser source (provided by LASERLINE Company Munich, Germany), and very similar results were achieved. This phenomenon can be attributed to two reasons. Firstly, the surface temperature of the prepreg was controlled in a closed-loop manner using the layup equipment, and the irradiation point temperature remained stable at the set value of 400 °C. Secondly, the amount of added powder particles was minimal, which had little effect on the thermal properties of the laminate.

The surface temperature of the P1 sample measured with the infrared thermal imager is shown in Figure 6a. Surface temperature data for the first, second, third, and fourth layers were recorded. The temperature rapidly increases with laser irradiation, and the heating rate and peak temperature are regulated with the laser heating controller. Since the processing parameters for each layer are exactly the same, the thermal histories are almost identical. Figure 6 also shows that the temperature before and after roller contact is 370 °C and 290 °C, respectively. This indicates that the resin remains in a molten state for approximately 1 s.

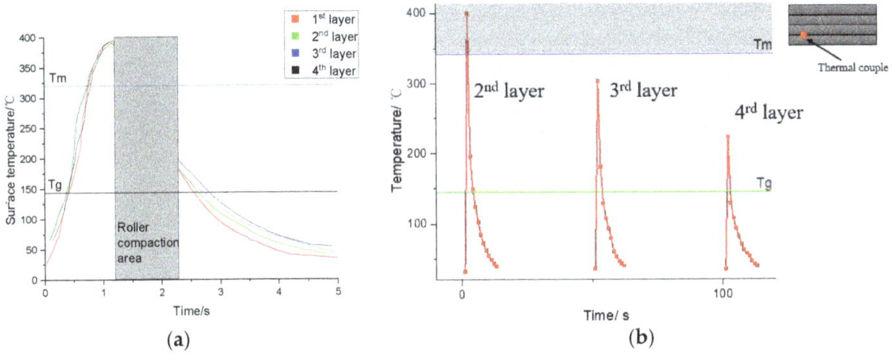

Figure 6. Thermal history of prepreg during layup. surface temperature (**a**); interface temperature (**b**).

Figure 6b shows the temperature at a spot between the first and second layers during the process obtained with a K-type thermocouple. As shown in Figure 6b, the thermal couple was attached to the first layer. When laying the second layer, the spot temperature matches the set temperature at 400 °C, higher than the Tm of PEEK(343 °C). When laying the third and fourth layers, laser heat conduction occurred downwards, causing the peak temperature at the spot to reach 315 °C and 230 °C, respectively. The transferred heat increased the local temperature above the resin's Tg point (251 °C).

Although the interlayer bonding time for in situ consolidation was very short, when laying subsequent layers, heat and pressure conduction occurred downwards, causing the substrate prepreg to undergo extra heating and compaction. The interlayer resin of the substrate prepreg cannot melt during this process, nor can it fill the interlayer voids. However, the additional heat and pressure facilitate the segment movement of the resin [29], which positively affects the interlayer bonding strength.

3.3. Wedge Peel Strength

The wedge peel strength of all samples was tested to assess the effect of coated powder on interlayer strength. The test results are shown in Figure 7. The unstable peeling force was induced at the beginning and end of a sample. This may be attributed to the change in the speed of the universal test machine and interlayer defects. Unstable data were excluded from the analysis, and the average wedge peel force of each group of samples was calculated accordingly.

The wedge peel strengths of the samples were calculated as $S = F/w$, where S, F, and w mean the wedge peel strength, wedge peel force, and sample width, respectively. To make the results more reliable, the width of every sample was measured and trimmed to 6.35 mm. Therefore, the defects on the edges of the samples did not affect the experimental results. The wedge peel strength of each experimental group is shown in Figure 8.

As shown in Figure 8, when the prepreg surface is uncoated, the peel strength is 2.6 N/mm. As the powder coating amount increases, the peel strength increases, reaching 3.4 ± 0.3 N/mm and 3.2 ± 0.4 N/mm, respectively. This indicates that a small amount of resin powder can melt at a laying speed of 100 mm/s, and the melted resin quickly flows to fill the interlayer voids, resulting in tighter interlayer bonding. However, when the average thickness of the coated powder increases to 21.6 g/m^2, the peel strength decreases. Due to the significant difference in laser absorption between resin particles and carbon fibers, carbon fibers primarily absorb the radiation heat of the laser beam. The heat is then transferred to the surrounding resin through thermal conduction. When the powder coating is too much, or the resin layer thickness is too thick, the energy transferred by thermal conduction is insufficient to melt all the coated resin. The unmelted resin particles instead greatly hinder interlayer bonding, resulting in lower peel strength and larger scatter values.

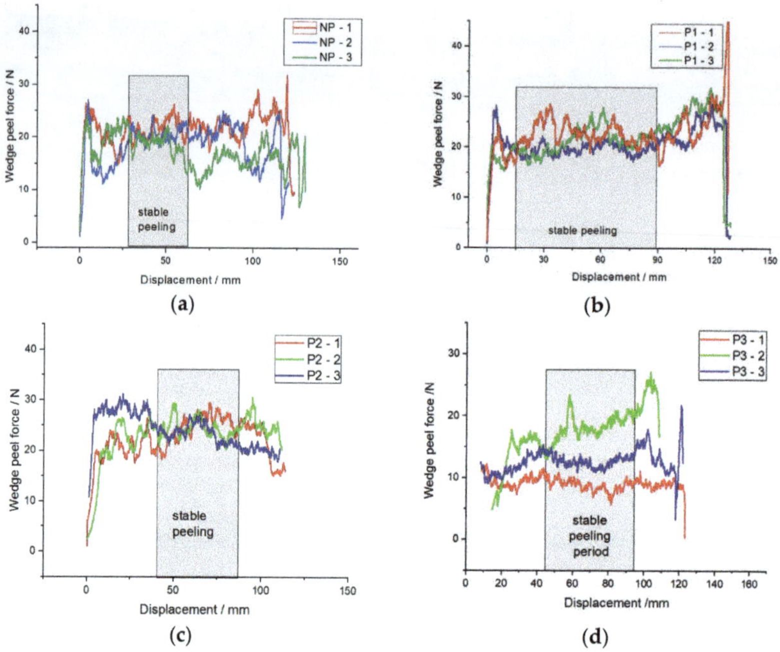

Figure 7. Force–displacement graphs for wedge peel test: NP (**a**), P1 (**b**), P2 (**c**), and P3 (**d**).

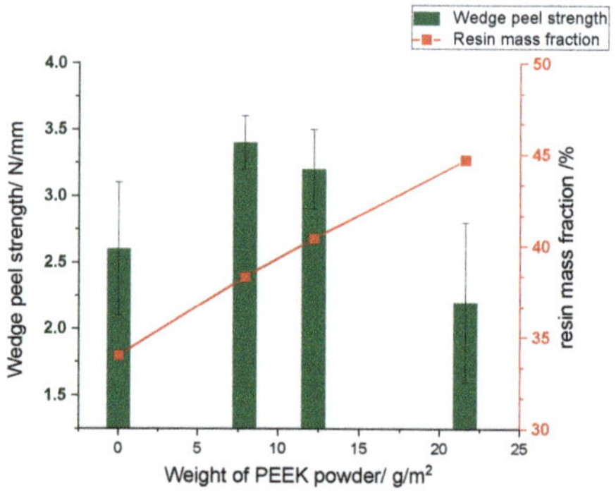

Figure 8. The effect of the weight of coated powder on wedge peel strength and resin mass fraction.

3.4. Porosity and Microstructure

The microscopic cross-sectional images of the specimens prepared under different experimental conditions are shown in Figure 9. And the effect of the weight of coated powder on porosity in shown in Figure 10.

Figure 9a depicts a specimen without surface powder coating, where noticeable voids are present. The porosity content of the sample is 7.2% as shown in Figure 10. Figure 9b,c show micrographs of specimens with 7.9 g/m^2 and 12.2 g/m^2 coated PEEK powder, and sample porosity is 3.4% and 2.1% respectively. In Figure 9b, regions with higher resin content can be observed. Despite the lower resin addition, there are still voids between the layers, but the dispersion of fibers and resin becomes relatively more uniform. This indicates that the added resin melts and flows during the layup process, which fills the interlayer voids under compaction force. At this point, the interlayer wedge peel strength is also the highest, reaching 3.3 N/mm. As shown in Figure 9c, as the amount of added powder increases to 12.2 g/m^2, a clear resin-rich area appears between the layers.. If the resin coating amount continues to increase and reaches 21.6 g/m^2, as shown in Figure 9d, voids begin to appear again in the interlayer resin region. According to Figure 10, sample porosity rises to 8.0%. This indicates that the resin is not fully melted at this point and that resin particles impede interlayer bonding.

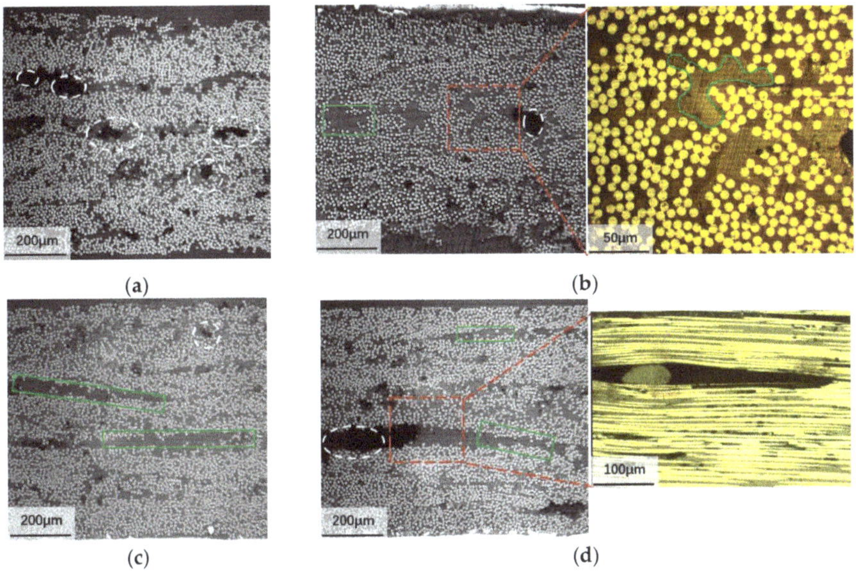

Figure 9. Microstructure of interleaved laminates: (**a**) NP, (**b**) P1, (**c**) P2, and (**d**) P3. White ellipses indicate voids and green rectangles indicate resin-rich areas.

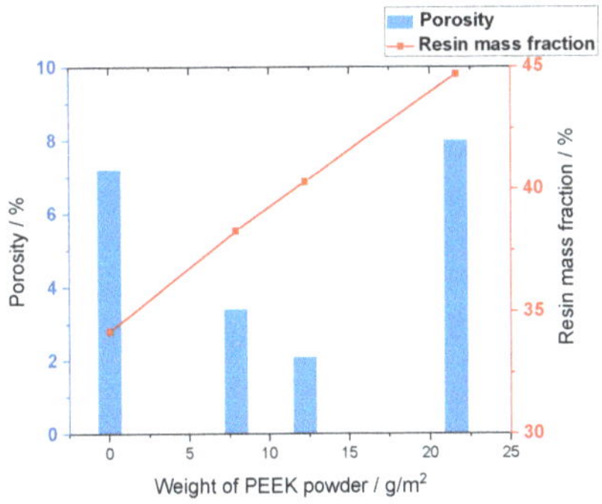

Figure 10. The effect of the weight of coated powder on porosity and resin mass fraction.

Upon zooming in on Figure 9b, it can be observed that the coated resin particles not only fill the interlayer voids after melting but also infiltrate along the voids toward the thickness direction of the prepreg. This is because the resin inside the prepreg experiences the hindrance effect of fibers, resulting in weaker flowability in the thickness direction, making it difficult to fill the interlayer voids [13]. However, the resin coated between layers exhibits strong fluidity, filling the interlayer voids and infiltrating "backward" into the prepreg. The added resin powder effectively reduced the porosity of the laminate.

If the powder coating is excessive, for example, exceeding 20 g/m^2, it becomes difficult to melt during the laying process. This results in longer defect areas along the interface. In Figure 9d, aggregated resin was found in the microscopic image of the specimen parallel to the fiber direction. Although the resin is heated and deformed into an oval shape, it does not fully melt. As a result, the surrounding fibers form "eye-shaped" defect areas. This discovery suggests that for resin interlayers thicker than 30 μm, achieving in situ consolidation by LAFP becomes impracticable.

The schematic diagram of interlayer void formation and the void filling of coated resin during layup are shown in Figure 11. There are some dry fibers and fiber-rich areas on the tape surface as depicted in Figure 11a. Laser heating cannot increase the resin content on the surface; instead, it tends to soften and roughen the surface. Due to the short processing time of high temperature and consolidation force, it is not possible to let resin flow completely [8], which causes many interlaminar voids and dry fibers, as depicted in Figure 11a. The added resin particles can rapidly melt during laser heating, flow under pressure, and fill the interlayer voids. This mechanism elucidates how powder coating can reduce the interlayer porosity and enhance peel strength, as depicted in Figure 11b. However, if excessive PEEK powder is added between the layers, it may not fully melt during the consolidation process. The unmelted PEEK particles are heated and compressed by compaction force, forming an "eye-shaped" defect area. This is shown in Figure 11c.

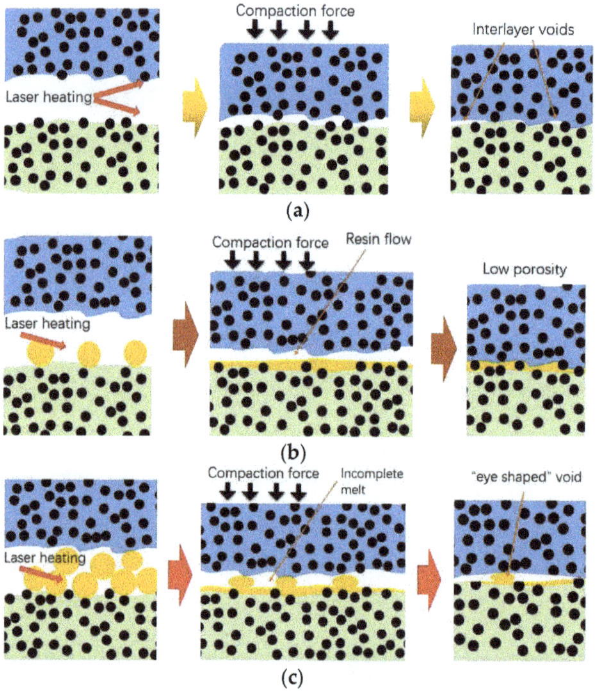

Figure 11. Schematic diagram of resin/fiber distribution: (**a**) during the LAFP process; (**b**) with coated PEEK powder; (**c**) excessive PEEK powder.

4. Conclusions

In this study, a novel approach involving the coating of fine PEEK powder on prepreg was introduced to enhance wedge peel strength and reduce interlaminar voids in CF/PEEK laminates consolidated in situ by laser-assisted fiber placement. Morphological analysis revealed a decrease in dry fiber and more uniform fiber–resin distribution. The thermal history during layup indicated a short bonding time as a dominant factor influencing void formation and limited interlayer strength. Furthermore, wedge peel strength significantly increased with the addition of a small quantity of resin, attributed to improved resin contact and reduced interlayer voids.

The successful implementation of this novel approach highlights its potential for improving the mechanical properties of thermoplastic composites manufactured by laser-assisted fiber placement. By optimizing resin powder coating levels, it is possible to achieve enhanced interlayer bonding strength while maintaining fiber-related mechanical properties. Future research could explore the applicability of this method across various thermoplastic composite systems and processing conditions, further advancing the field of additive manufacturing in aerospace.

Author Contributions: Conceptualization, methodology, and writing—original draft preparation, R.W.; validation, R.W. and E.X.; resources, R.W. and L.W.; project administration, L.W. All authors have read and agreed to the published version of the manuscript.

Funding: This research received no external funding.

Institutional Review Board Statement: Not applicable.

Informed Consent Statement: Not applicable.

Data Availability Statement: The data presented in this study are available upon request from the corresponding author.

Conflicts of Interest: The authors declare no conflicts of interest.

·erences

Mills, A. Automation of carbon fibre preform manufacture for affordable aerospace applications. *Compos. Part A Appl. Sci. Manuf.* **2001**, *32*, 955–962. [CrossRef]
Bandaru, A.K.; Clancy, G.; Peeters, D.; O'Higgins, R.M.; Weaver, P.M. Properties of a thermoplastic composite skin-stiffener interface in a stiffened structure manufactured by laser-assisted tape placement with in situ consolidation. *Compos. Struct.* **2019**, *214*, 123–131. [CrossRef]
Penumakala, P.K.; Santo, J.; Thomas, A. A critical review on the fused deposition modeling of thermoplastic polymer composites. *Compos. Part B Eng.* **2020**, *201*, 108336. [CrossRef]
Qureshi, Z.; Swait, T.; Scaife, R.; El-Dessouky, H.M. In situ consolidation of thermoplastic prepreg tape using automated tape placement technology: Potential and possibilities. *Compos. Part B Eng.* **2014**, *66*, 255–267. [CrossRef]
Comer, A.J.; Ray, D.; Obande, W.O.; Jones, D.; Lyons, J.; Rosca, I.; O' Higgins, R.M.; McCarthy, M.A. Mechanical characterisation of carbon fibre–PEEK manufactured by laser-assisted automated-tape-placement and autoclave. *Compos. Part A Appl. Sci. Manuf.* **2015**, *69*, 10–20. [CrossRef]
Stokes-Griffin, C.M.; Compston, P. The effect of processing temperature and placement rate on the short beam strength of carbon fibre–PEEK manufactured using a laser tape placement process. *Compos. Part A Appl. Sci. Manuf.* **2015**, *78*, 274–283. [CrossRef]
Chen, J.; Fu, K.; Li, Y. Understanding processing parameter effects for carbon fibre reinforced thermoplastic composites manufactured by laser-assisted automated fibre placement (AFP). *Compos. Part A Appl. Sci. Manuf.* **2021**, *140*, 106160. [CrossRef]
Liu, H.; Li, Y.; Huan, D.; Wang, W.; Li, Y.; Li, L. Influence of Curvature Feature on Laser Heating during Tape Placement Process for Carbon Fiber Reinforced Polyether Ether Ketone Composite. *Polymers* **2023**, *15*, 289. [CrossRef]
Song, Q.; Liu, W.; Chen, J.; Zhao, D.; Yi, C.; Liu, R.; Geng, Y.; Yang, Y.; Zheng, Y.; Yuan, Y. Research on void dynamics during in situ consolidation of CF/high-performance thermoplastic composite. *Polymers* **2022**, *14*, 1401. [CrossRef]
Miao, Q.; Dai, Z.; Ma, G.; Niu, F.; Wu, D. Effect of consolidation force on interlaminar shear strength of CF/PEEK laminates manufactured by laser-assisted forming. *Compos. Struct.* **2021**, *266*, 113779. [CrossRef]
Lee, W.I.; Springer, G.S. A model of the manufacturing process of thermoplastic matrix composites. *J. Compos. Mater.* **1987**, *21*, 1017–1055. [CrossRef]
Kim, Y.H.; Wool, R.P. A theory of healing at a polymer-polymer interface. *Macromolecules* **1983**, *16*, 1115–1120. [CrossRef]
Çelik, O.; Peeters, D.; Dransfeld, C.; Teuwen, J. Intimate contact development during laser assisted fiber placement: Microstructure and effect of process parameters. *Compos. Part A Appl. Sci. Manuf.* **2020**, *134*, 105888. [CrossRef]
Çelik, O.; Choudhary, A.; Peeters, D.; Teuwen, J.; Dransfeld, C. Deconsolidation of thermoplastic prepreg tapes during rapid laser heating. *Compos. Part A Appl. Sci. Manuf.* **2021**, *149*, 106575. [CrossRef]
Slange, T.K.; Warnet, L.L.; Grouve, W.J.B.; Akkerman, R. Deconsolidation of C/PEEK blanks: On the role of prepreg, blank manufacturing method and conditioning. *Compos. Part A Appl. Sci. Manuf.* **2018**, *113*, 189–199. [CrossRef]

16. Hulcher, A.B.; Banks, W.I., III; Pipes, R.B.; Tiwari, S.N.; Cano, R.J.; Johnston, N.J. Automated fiber placement of PEEK/IM composites with film interleaf layers. In Proceedings of the 46th International SAMPE Symposium and Exhibition, Covina, C USA, 6–10 May 2001; pp. 1998–2012.
17. Miao, Q.; Dai, Z.; Ma, G.; Niu, F.; Wu, D. CF/PEEK interleaved laminates with PEEK film interleaving manufactured laser-assisted forming: Microstructure and interlaminar shear strength. *Compos. Part A Appl. Sci. Manuf.* **2023**, *172*, 1075 [CrossRef]
18. Wang, X.; Chen, C.; Hu, S.; Chen, Z.; Jiang, W.; Shen, G.; Huang, Z.; Zhou, H. In-situ infrared annealing for laser-assist automated fiber placement to enhance interlaminar properties without sacrificing laydown efficiency. *Compos. Part A Appl. S Manuf.* **2024**, *183*, 108214. [CrossRef]
19. Chanteli, A.; Bandaru, A.K.; Peeters, D.; O'Higgins, R.M.; Weaver, P.M. Influence of repass treatment on carbon fibre-reinforc PEEK composites manufactured using laser-assisted automatic tape placement. *Compos. Struct.* **2020**, *248*, 112539. [CrossRef]
20. Kim, J.W.; Lee, J.S. Influence of interleaved films on the mechanical properties of carbon fiber fabric/polypropylene thermoplas composites. *Materials* **2016**, *9*, 344. [CrossRef]
21. de Souza, G.; Tarpani, J.R. Interleaving CFRP and GFRP with a thermoplastic ionomer: The effect on bending properties. *Ap Compos. Mater.* **2021**, *28*, 559–572. [CrossRef]
22. Ourahmoune, R.; Salvia, M.; Mathia, T.G.; Mesrati, N. Surface morphology and wettability of sandblasted PEEK and composites. *Scanning J. Scanning Microsc.* **2014**, *36*, 64–75. [CrossRef] [PubMed]
23. Abbasnezhad, N.; Khavandi, A.; Fitoussi, J.; Arabi, H.; Shirinbayan, M.; Tcharkhtchi, A. Influence of loading conditions on t overall mechanical behavior of polyether-ether-ketone (PEEK). *Int. J. Fatigue* **2018**, *109*, 83–92. [CrossRef]
24. Zhang, C.; Duan, Y.; Xiao, H.; Wang, B.; Ming, Y.; Zhu, Y.; Zhang, F. The effects of processing parameters on the wedge pe strength of CF/PEEK laminates manufactured using a laser tape placement process. *Int. J. Adv. Manuf. Technol.* **2022**, 1 7251–7262. [CrossRef]
25. *ASTM D2344*; Standard Test Method for Short-Beam Strength of Polymer Matrix Composite Materials and Their Lamina ASTM: West Conshohocken, PA, USA, 2016.
26. *ASTM D5528*; Standard Test Method for Mode I Interlaminar Fracture Toughness of Unidirectional Fiber-Reinforced Polym Matrix Composites. ASTM: West Conshohocken, PA, USA, 2013.
27. Stokes-Griffin, C.M.; Kollmannsberger, A.; Compston, P.; Drechsler, K. The effect of processing temperature on wedge pe strength of CF/PA 6 laminates manufactured in a laser tape placement process. *Compos. Part A Appl. Sci. Manuf.* **2019**, *121*, 84– [CrossRef]
28. Johnston, N.J.; Belvin, H.L.; Cano, R.J.; Marchello, J.M.; Hulcher, A.B. A prototype research laboratory for automated fabricati of high performance composites. In Proceedings of the 12th International Conference on Composite Materials ICCM12, Pa France, 5–9 July 1999.
29. Stokes-Griffin, C.M.; Compston, P. Investigation of sub-melt temperature bonding of carbon-fibre/PEEK in an automated la tape placement process. *Compos. Part A Appl. Sci. Manuf.* **2016**, *84*, 17–25. [CrossRef]

coatings

icle

licrostructure, Hardness, Wear Resistance, and Corrosion esistance of As-Cast and Laser-Deposited ›CoNiCrAl0.8Cu0.5Si0.5 High Entropy Alloy

ohu Ji [1,2,3,*], Lihua Zhou [3] and Heng Li [3]

[1] School of Mechanical and Vehicle Engineering, West Anhui University, Yueliangdao Road, No. 1, Lu'an 237010, China
[2] Innovation Platform of High-Performance Complex Manufacturing Intelligent Decision and Control, West Anhui University, Yueliangdao Road, No. 1, Lu'an 237010, China
[3] Anhui Province Key Lab of Aerospace Structural Parts Forming Technology and Equipment, Hefei University of Technology, No. 193 Tunxi Road, Baohe District, Hefei 230009, China; zhoulihua@hfut.edu.cn (L.Z.); liheng0205@hfut.edu.cn (H.L.)
* Correspondence: 04000169@wxc.edu.cn; Tel.: +86-18255171038

Abstract: FeCoNiCrAl0.8Cu0.5Si0.5 high-entropy alloys were fabricated using vacuum induction melting and laser deposition processes, followed by a comparison of the structural and mechanical properties of two distinct sample types. The as-cast FeCoNiCrAl0.8Cu0.5Si0.5 alloy is comprised of BCC1, BCC2, and Cr3Si phases, while the laser-deposited alloy primarily features BCC1 and BCC2 phases. Microstructural analysis revealed that the as-cast alloy exhibits a dendritic morphology with secondary dendritic arms and densely packed grains, and the laser-deposited alloy displays a dendritic structure without the formation of granular interdendritic regions. For mechanical properties, the as-cast FeCoNiCrAl0.8Cu0.5Si0.5 alloy demonstrated higher hardness than the as-deposited alloy, with values of 586 HV0.2 and 557 HV0.2, respectively. The wear rate for the as-cast alloy was observed at 3.5×10^{-7} mm^3/Nm, with abrasive wear being the primary wear mechanism. Conversely, the as-deposited alloy had a wear rate of 9.0×10^{-7} mm^3/Nm, characterized by adhesive wear. The cast alloy exhibited an icorr of 4.062 μA·cm^{-2}, with pitting as the form of corrosion. The laser-deposited alloy showed an icorr of 3.621 μA·cm^{-2}, with both pitting and intergranular corrosion observed. The laser-deposited alloy demonstrated improved corrosion resistance. The investigation of their microstructure and mechanical properties demonstrates the application potential of FeCoNiCrAl0.8Cu0.5Si0.5 alloys in scenarios requiring high hardness and enhanced wear resistance.

Keywords: FeCoNiCrAl0.8Cu0.5Si0.5; high-entropy alloy; laser deposition; microstructure; wear resistance; corrosion resistance

1. Introduction

Multicomponent alloys have transcended the constraints of traditional alloy design principles, expanding their compositional range into the central regions of the multicomponent phase diagram [1–3]. High-entropy alloys (HEAs), a subset of multicomponent alloys, are typically composed of four or more alloying elements in equiatomic or near-equiatomic ratios. These alloys exhibit unique microstructural characteristics and noteworthy mechanical and physical properties, positioning them as potential substitutes for some traditional alloys and lightweight materials across various industries, including aerospace, biomedical, and marine [4–7].

The initial exploration of HEA was driven by the high-entropy effect, aiming to inhibit the formation of intermetallic compounds through the combination of multiple elements [8,9]. This approach enabled the creation of single-phase alloy structures, such as the FCC and

BCC structures found in CoCrCuFeNi and AlCoCrFeNi HEAs, respectively [10–14]. The CoCrFeMnNi HEA, with its characteristic FCC structure, boasts high thermodynamic stability, superb malleability and machinability, exceptional fracture toughness, and an elevated limit for low-temperature damage resistance. AlCoCrFeNi HEA, which possesses a BCC structure, is distinguished by its significant strength. Although FCC alloys exhibit high toughness and single-phase BCC alloys demonstrate high strength, there remains a need to develop alloys that can simultaneously meet the requirements for high toughness and high strength. Regarding the corrosion of HEAs, the principal metal elements have a significant impact. The Cr element is typically used as a solute in the synthesis of stainless-steel materials. It enhances the alloy by forming a dense oxide film, as seen in martensitic stainless steels (such as 1Cr13, 2Cr13Ni2, 3Cr13, and 4Cr13) and austenitic stainless steels (alloy steels containing elements like Cr-Ni and Cr-Ni-Mn). The element promotes the formation of the BCC in CoCrFeNi alloys, causing the primary phase to shift from FCC to a dual-phase structure of FCC and BCC. This shift leads to galvanic corrosion between different phases within the alloy, reducing its overall corrosion resistance. The Co promotes the formation of the phase in HEAs. As the proportion of the FCC phase increases and the BCC phase decreases, the toughness of the alloy improves, which can significantly expand the application of HEAs in specific service environments. However, an increase in Co content in HEAs reduces the proportion of other corrosion-resistant elements, thereby affecting the overall corrosion resistance of the alloy. The mixing enthalpy of the Cu element is positive, indicating weak bonding with most other elements. This often results in segregation in the interdendritic regions. Galvanic corrosion occurs between Cu-rich and Cu-poor phases, with the Cu-rich phase preferentially dissolving during metal corrosion. Therefore, HEAs containing Cu generally exhibit poor corrosion resistance.

With advancements in HEAs, performance enhancements have been realized through adjustments in manufacturing techniques and changes in elemental compositions. The AlCoCrFeNi2.1 eutectic HEA possesses FCC and BCC structures [15,16]. Combined with a suitable casting process, it achieves a balanced interaction between strength and ductility. Traditional casting methods, such as directional solidification and vacuum induction melting, are commonly used for producing eutectic HEAs. These methods ensure process reliability and reduced defect rates. However, the as-cast eutectic HEAs typically have coarse microstructures, with lamellae thicknesses in the micrometer range, posing challenges to strengthening efforts [17–19]. Additionally, realizing the desired properties in eutectic HEA components often requires further thermomechanical treatments [16,20,21]. Hsu et al. studied the friction and wear performance of the AlCoCrFexMo0.5Ni high-entropy alloy system [22]. The trend indicates that hardness is inversely proportional to the coefficient of friction, and the higher the hardness of the alloy, the better its wear resistance. The wear resistance of body-centered cubic phase alloys is superior to that of face-centered cubic phase alloys. Additionally, the more significant the solid solution strengthening effect, the better the wear resistance of the alloy. Chuang et al. studied the wear resistance of AlxCo 1.5CrFeNi1.5Ti series high-entropy alloys and found that Co1.5CrFeNi1.5Ti and Al0.2Co 1.5CrFeNi1.5Ti alloys exhibit excellent wear resistance.

Adjusting the content of principal elements can also enhance alloy performance. In the AlxCrFeCoNi HEA series (where x ranges from 0 to 2.0 in molar ratio), the increase in Al content leads to a transition in the alloy's microstructure: from single-phase FCC at $x = 0$ to 0.4 to dual-phase FCC + BCC at $x = 0.5$ to 0.8, and finally to single-phase BCC at $x = 0.85$ to 2.0 [23]. Alloys with a single-phase FCC structure exhibit good ductility but lower strength. For example, the Al0.3CrFeCoNi HEA has a tensile elongation of approximately 47%, with a yield strength of approximately 259 MPa. On the other hand, alloys with a single-phase BCC structure display higher strength but reduced ductility. Alloys with a dual-phase FCC + BCC structure offer a favorable combination of high strength and ductility.

Metallurgical theories indicate that alloy properties are influenced not only by manipulating metal elements like Ti, Cr, and Al but also by the incorporation of non-metal elements, which are key in the microstructure and performance of high-entropy alloys

(HEAs). Research led by Zhang et al. showed that adding boron to FeCoCrNi HEAs triggers the precipitation of the Fe3B, which improves the hardness and wear resistance. In marine environments, the corrosion resistance of bridge steel related to Si content initially increases and then decreases, suggesting a complex mechanism of Si's impact on corrosion properties. HEAs, characterized by multiple components, high entropy of mixing, significant lattice distortion, and slow diffusion, are also affected by the addition of non-metal elements like Si, which influence their mechanical properties. Thus, considering non-metal elements such as Si is crucial in the design and research of HEAs to optimize their microstructure and performance.

Recently, additive manufacturing has provided a new avenue for producing HEA components with complex geometrical structures [24–26]. Among these, laser deposition, a powder-fed additive manufacturing technique, operates on the principle of using a high-energy laser beam for point-by-point, layer-by-layer scanning to rapidly melt and solidify the powder, thereby directly constructing integrated three-dimensional metal components. Combining the advantages of HEAs and this technology holds the promise of overcoming manufacturing bottlenecks for high-performance complex structural components. Current advancements in optimizing laser process parameters have enabled the printing of nearly fully dense HEAs, with room-temperature mechanical properties significantly surpassing those of cast alloys [27,28]. Further research indicates that the high-temperature gradients and cooling rates during the laser deposition process are conducive to the formation of alloys with excellent microstructural morphology, significantly enhancing the interphase strengthening effect and verifying the feasibility of tailoring the mechanical properties of HEAs [29,30].

This study leverages the theory of forming solid solution alloys by adjusting metal element ratios under the high-entropy effect and further regulating alloy mechanical properties by adding non-metal elements. We also designed the FeCoNiCrAl0.8Cu0.5Si0.5 HEA structure. By comparing HEAs in both laser-deposited and as-cast conditions, we investigate the influence of different processing techniques on the microstructure and properties.

2. Materials and Methods

The experiment prepared FeCoNiCrAl0.8Cu0.5Si0.5 HEA samples using a vacuum induction melting furnace and a Rofin DC050 laser deposition manufacturing system. For cast samples, raw materials in pure forms of Si, Cr, Cu, Co, Ni, Fe, and Al were used to prepare cast samples. The vacuum induction furnace melts these materials under a high-purity argon atmosphere, preventing contamination and oxidation. The induction process generates heat through electromagnetic induction, allowing for uniform heating and melting of the materials. This melting was repeated four times to ensure a homogeneous composition. The molten alloy was then cast into high-entropy alloy button samples using a copper mold in an argon atmosphere.

Laser deposition was performed using the coaxial powder feeding method. This method works by directing a high-power laser beam and a stream of metal powder simultaneously onto a substrate. The laser melts the substrate and the powder, which is fed through a nozzle aligned with the laser beam. As the laser moves, the molten material solidifies, forming a new layer. Spherical powders of Fe, Co, Ni, Al, Cr, Cu, and Si, provided by Changsha Tian Jiu Powder Co. (Changsha, China), with a purity greater than 99.5% and a particle size of 200–325 mesh, were selected. The powders were mixed using a planetary ball mill, and argon gas was added to the mixing chamber to prevent oxidation of the elements during the mixing process. Our research team conducted systematic process experiments on the Al0.8CrFeCoNiCu0.5 Hê, ultimately determining the optimal laser deposition parameters for achieving the best-forming quality, which are as follows [31]: a beam diameter of 1.2 mm; a laser power of 1850 W; a powder feed rate of 5.6 g/min; a scanning speed of 120 mm/min; a duty cycle of 70%; a pulse frequency of 50 Hz; an overlap rate of 30%; and a carrier gas flow rate of 5 L/min.

The HEA was characterized using an Empyrean Panalytical XRD, scanning at 4 (°)/min within a range of 20° to 90°. This scanning range was selected to ensure comprehensive coverage of the most relevant diffraction peaks, providing accurate phase identification. An SEM and EDS (JSM-6510,Tokyo, Japan) were utilized to observe the samples' microstructure and wear morphologies. These instruments were chosen due to their high resolution, enabling detailed analysis of the microstructure to correlate with wear performance.

Hardness was assessed with an HV-1000 microhardness teste (Shenyang, China), applying a 1000 g load with a 10 s dwell time. This specific load and dwell time were chosen to ensure repeatable measurements while minimizing potential indentation size effects that could skew the hardness values.

The corrosion test for the samples was recorded using a Nova2 electrochemical workstation provided by Metrohm (Herisau, Switzerland) The equipment was selected due to its high sensitivity and accuracy, which is crucial for capturing fine variations in electrochemical behavior. Analysis and computational tasks related to the data were conducted using ZView software (Scribner Associates Inc., Southern Pines, NC, USA).

The setup for the experiments featured a tri-electrode system, where the HEA was designated as the working electrode. A saturated calomel electrode was chosen as the reference due to its stable and well-defined potential, while a platinum sheet electrode was used as the auxiliary electrode for its high conductivity and resistance to corrosion. Electrochemical impedance spectroscopy (EIS) measures the response by applying a small sinusoidal perturbation (potential or current) and calculating the impedance at each frequency. Therefore, EIS can detect minute changes in the electrode surface state and relevant electrochemical processes at the electrode/electrolyte interface without altering the surface state.

An MM200-type machine (Jinan Hengxu Testing Machine Co., Ltd., Changzhou, China) was used to assess the samples' wear test. Testing conditions were set to room temperature and 20% relative humidity. A GCr15 steel ball was selected as the friction pair due to its standard use in wear testing, ensuring comparability with the existing literature. Testing was performed for 30 min under specified loads, which were carefully chosen to mimic practical applications. The ball's velocity, hardness (61 HRC), diameter (50 mm), and thickness (10 mm) were noted. Wear resistance was calculated using the formula:

$$\omega = \frac{V_{loss}}{L \times N}$$

$$L = 2\pi R v t$$

$$V_{loss} = B\left[\frac{\pi R^2}{180}\arcsin\left(\frac{b}{2R}\right) - \frac{b}{2}\sqrt{R^2 - \frac{b^2}{4}}\right]$$

where V_{loss} represents the wear volume, B is the length of the wear track (mm), b is the width of the wear track (mm), v is the wheel speed (r/min), and L is the total sliding distance of the wear block on the sample surface (mm).

3. Results and Discussion

3.1. Phase Analysis

Figure 1 presents the XRD spectra of both cast and laser-deposited FeCoNiCrAl0.8Cu0.5 Si0.5 samples. It is observed that the laser-deposited alloy exhibits BCC1 and BCC2 while the cast alloy shows BCC1, BCC2, and Cr3Si phases. This indicates that the manufacturing process influences the phase precipitation behavior of the FeCoNiCrAl0.8Cu0.5Si0.5 HEA. According to Gibbs' phase rule [32], a seven-component alloy system can potentially form up to eight phases. In the FeCoNiCrAl0.8Cu0.5Si0.5 HEA, the actual number of phases is significantly less than the theoretical maximum. This discrepancy arises because the diversity of elements and the increased atomic percentage of each element in the alloy elevate the system's mixing entropy. Although the as-cast FeCoNiCrAl0.8Cu0.5Si0.5 HEA produces Cr3Si, the significantly reduced Gibbs free energy limits the formation of additional phases, still resulting in an alloy predominantly composed of BCC phases.

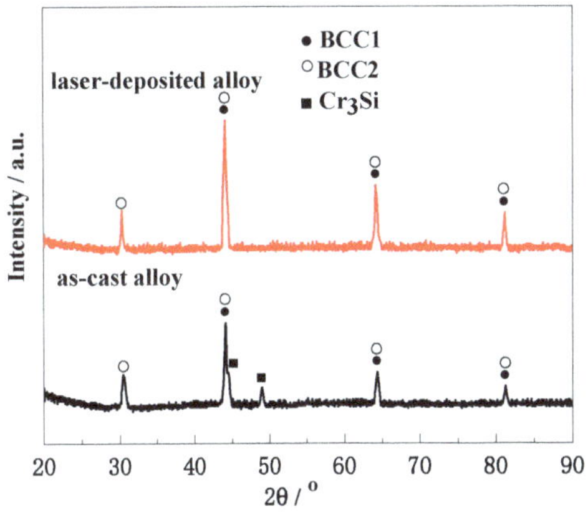

Figure 1. The XRD of FeCoNiCrAl0.8Cu0.5Si0.5 HEA.

Furthermore, the presence of Cr3Si in the cast alloy suggests that the silicon content exceeds the alloy's solubility limit. The formation of Cr3Si could be attributed to the high melting point of chromium (1907 °C), which may preferentially bond with silicon during the alloy solidification process to form Cr3Si compounds. Additionally, despite the differences in lattice types between chromium (BCC) and silicon (DC), compared to aluminum (FCC), iron (FCC), cobalt (FCC), and nickel (FCC), the lattice type mismatch between silicon and chromium is relatively minor. Since the added silicon content exceeds the solubility limit in the alloy, it forms compounds with Cr. The absence of Cr3Si in the laser-deposited FeCoNiCrAl0.8Cu0.5Si0.5 HEA can be speculated to relate to the rapid heating and cooling during the laser deposition process, which provides inadequate time for silicon compounds to precipitate.

To identify the key elements influencing the phase formation of the FeCoNiCrAl0.8Cu0.5 Si0.5 HEA, phase equilibrium information calculations were conducted. ThermoCalc software (version 2022a) was utilized, incorporating the latest high-entropy alloy database TCHEA6. Figure 2 illustrates the calculated equilibrium solidification of the FeCoNi CrAl0.8Cu0.5Si0.5 HEA. The data suggest that the alloy solidifies in the BCC phase, with the liquidus temperature at 1470 °C and the solidus temperature at 1276 °C. The Cr3Si phase emerges around 1136 °C, with its volume fraction increasing as the temperature decreases. Concurrently, the BCC2 phase appears near this temperature. When the temperature drops to 1115 °C, the FCC phase emerges, followed shortly by the FCC2 phase.

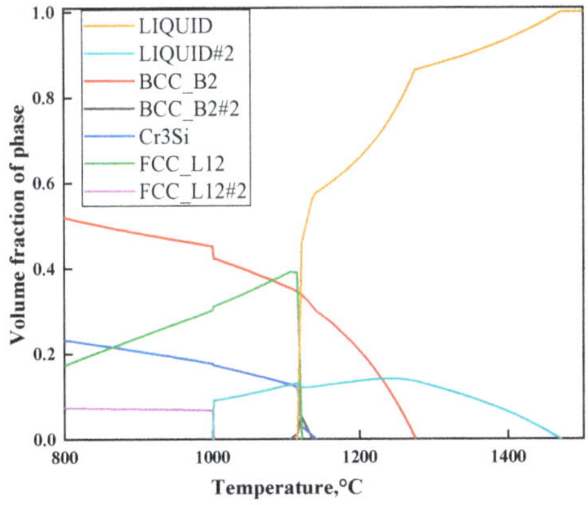

Figure 2. Phase equilibrium information of HEA predicted by ThermoCalc software.

3.2. *Microstructure Analysis*

Figure 3 shows the surface composition maps of as-cast and laser-deposited FeCoNiCrAl0.8Cu0.5Si0.5 HEAs, while Table 1 provides a comparison of the compositions. It can be observed that the contents of Fe, Co, Ni, Cr, Al, and Cu are close to the nominal composition of the alloy, while the Si content is slightly lower. Comparing the compositions of the cast and deposited alloys reveals that the Fe content is higher in the deposited alloy, while the Cu and Si contents are lower. This discrepancy may be due to the loss of Cu and Si during the high-temperature heating in the laser deposition process.

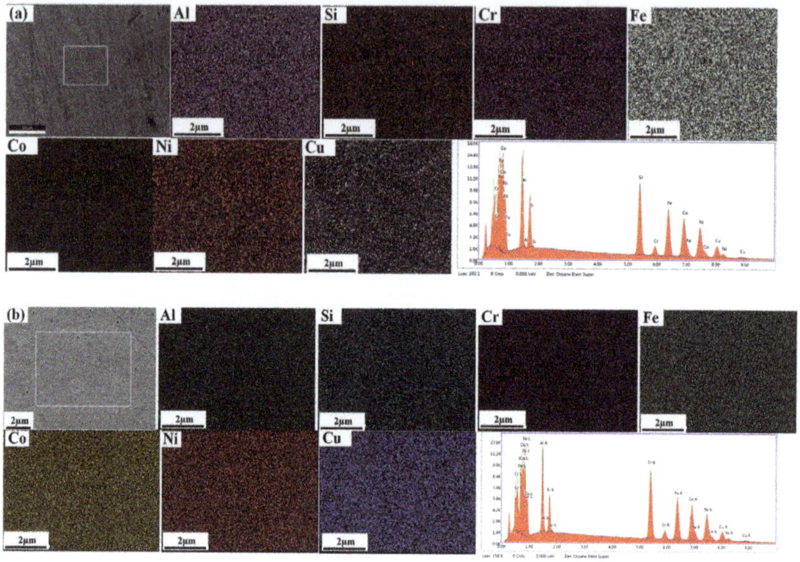

Figure 3. Surface composition maps of FeCoNiCrAl0.8Cu0.5Si0.5 HEAs: (**a**) as-cast alloy; (**b**) laser-deposited alloy.

Table 1. Composition of as-cast and laser-deposited FeCoNiCrAl0.8Cu0.5Si0.5 HEAs.

Element (Atomic %)	Al	Cr	Fe	Co	Ni	Cu	Si
As-cast alloy	13.85	18.94	18.23	17.20	17.04	8.24	6.50
Laer-deposited alloy	13.77	18.97	20.79	17.79	17.26	7.56	4.92

The microstructure of FeCoNiCrAl0.8Cu0.5Si0.5 HEAs was analyzed. Figure 4a depicts the microstructural morphology of the as-cast alloy. It is observed that the HEA exhibits a typical dendritic structure, with the formation of secondary dendrite arms. Figure 4b shows a magnified view of the dendritic arms, where a dense granular structure on the surface of the dendrites can be identified. The formation of secondary dendrite arms is generally attributed to solute segregation during the solidification process [33]. Figure 4c,d presents the morphology of laser-deposited alloy, revealing a granular morphology without the formation of secondary dendrite arms or granular structures. In Wang's study [34] on the NixCo0.6Fe0.2CrySizAlTi0.2 HEA, annealing the alloy at 800 °C for 10 h resulted in the precipitation of the Cr3Si hard phase. In Hsu's research [22] on NiCo0.6Fe0.2Cr1.5SiAlTi0.2 and NiCo0.6Fe0.2Cr1.3SiAl, the prepared alloys exhibited BCC, FCC, and Cr3Si structures. This indicates that in transition metal high-entropy alloys, Si tends to form the Cr3Si structure with Cr under certain processing conditions.

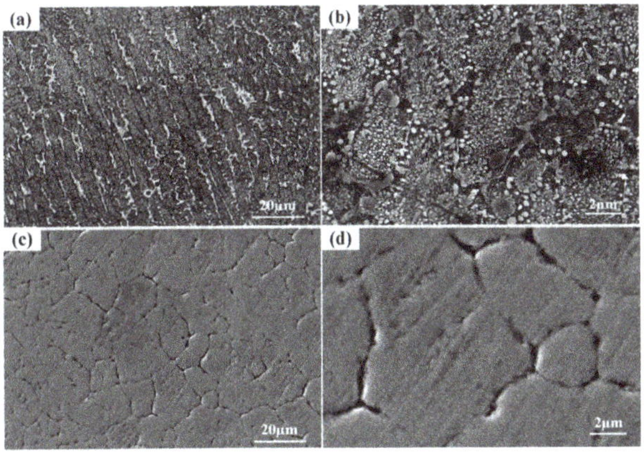

Figure 4. The microstructure of the FeCoNiCrAl0.8Cu0.5Si0.5 HEA: (**a**) low-magnification SEM of as-cast alloy; (**b**) high-magnification SEM of as-cast alloy; (**c**) low-magnification SEM of laser-deposited alloy; (**d**) high-magnification SEM of laser-deposited alloy.

A compositional scan of the laser-deposited alloy, as depicted in Figure 5, revealed the interdendritic segregation of Al, Ni, and Cu elements, while Si is uniformly distributed throughout the structure. XRD analysis indicates the alloy consists of two BCC phases, suggesting the interdendritic region is likely composed of an Al-Ni-rich BCC phase. Solidification theory explains that secondary dendritic arms arise from solute segregation during the alloy's solidification process [35]. With more than five elements mixed in the alloy, solute segregation can cause the alloy to deviate from its nominal composition, reducing local mixing entropy and failing to prevent the formation of high-enthalpy intermetallic compounds. For high-entropy alloys containing Si, the size difference in Si atoms facilitates a higher diffusion rate among elements, leading to easier compositional undercooling at the solid–liquid interface front under lower solidification rates. Additionally, the negative mixing enthalpy enhances the bonding between elements, making regions with a high Si content more prone to forming high-enthalpy intermetallic compounds during solidification. However, with sufficiently high cooling rates, the rapidly moving solid–liquid interface can more readily trap solutes, leading to a tendency to form metastable supersaturated solid solutions upon solidification. Therefore, compared to FeCoNiCrAl0.8Cu0.5Si0.5 alloys prepared using vacuum arc melting, rapid cooling techniques favor the formation of HEAs with simpler phase structures.

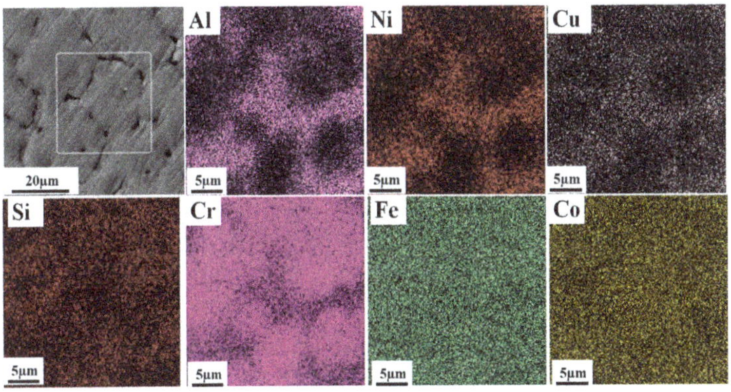

Figure 5. Compositional scan of the laser-deposited FeCoNiCrAl0.8Cu0.5Si0.5 HEA.

3.3. Hardness

Hardness tests were conducted on the FeCoNiCrAl0.8Cu0.5Si0.5 HEA, revealing that the hardness values for the cast and laser-deposited alloys were 586 HV0.2 and 557 HV0.2, respectively, with the laser-deposited alloy exhibiting lower hardness compared to the cast alloy. The absence of the Cr3Si hard phase in the laser-deposited alloy, which is present in the cast alloy, contributes to the increased hardness of the cast alloy. Compared to conventional steels, both as-cast and laser-deposited FeCoNiCrAl0.8Cu0.5Si0.5 HEAs show significantly higher hardness values than traditional steel materials, such as the 304 stainless steel (185 HV0.2) [36].

3.4. Wear Resistance

Wear testing was performed on both as-cast and laser-deposited FeCoNiCrAl0.8Cu0.5 Si0.5 HEAs. The as-cast alloy exhibited a scratch length, scratch width, and wear rate of 6.2 mm, 2.1 mm, and 3.5×10^{-7} $mm^3/(N{\cdot}m)$, respectively. The laser-deposited alloy displayed a scratch length, scratch width, and wear rate of 6.4 mm, 2.4 mm, and 9.0×10^{-7} mm^3, respectively. These results indicate that the as-cast alloy exhibits enhanced wear resistance relative to the laser-deposited alloy.

Figure 6 presents the wear surfaces of the FeCoNiCrAl0.8Cu0.5Si0.5 HEA. On the as-cast alloy surface, distinct plowing grooves are observable, indicative of abrasive wear mechanisms, as depicted in Figure 6a. The surface of the laser-deposited alloy exhibits adhesion, characteristic of adhesive wear, as shown in Figure 6b. The microstructure of the as-cast FeCoNiCrAl0.8Cu0.5Si0.5 HEA reveals the presence of secondary dendrites and particulate phases. Considering the wear morphology, it can be inferred that during the wear process, under the action of shear forces, the asperities of the counterface shear off the particulate phases from the material's surface. Under compressive forces, these particulate phases are driven into the matrix surface. As wear progresses, the asperities of the counterface repeatedly rub against the alloy surface, ultimately forming grooves. Moreover, due to the matrix's higher hardness, no adhesive layer is formed. The surface of friction samples from laser-deposited alloy exhibits features of adhesive wear. Under load, the asperities of the counterface cold weld to the surface of the sample, resulting in spalling pits upon detachment. Thus, the different processing methods significantly affect the FeCoNiCrAl0.8Cu0.5Si0.5 HEA's microstructure, ultimately influencing the alloy's wear behavior.

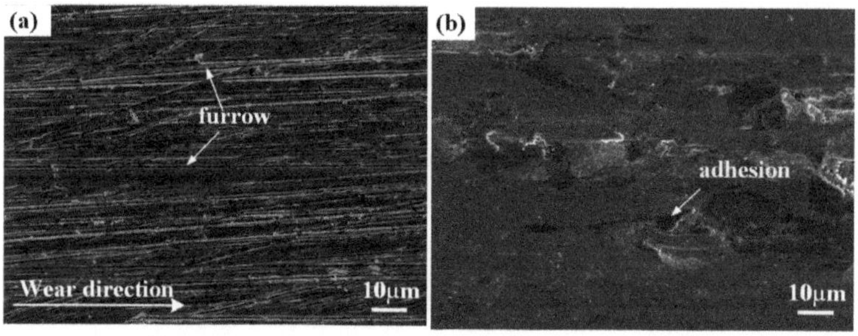

Figure 6. Wear morphology of FeCoNiCrAl0.8Cu0.5Si0.5 HEA: (**a**) as-cast alloy; (**b**) laser-deposited alloy.

3.5. Corrosion Resistance

Figure 7 presents the dynamic potential polarization curves and electrochemical impedance spectra (EIS) for two types of samples in a 3.5% NaCl solution. Table 2 summarizes fitting results for the case illustrated in Figure 7a. The corrosion current density (i_{corr}) for the cast alloy is 4.062 $\mu A{\cdot}cm^{-2}$, which is higher than that of the laser-deposited alloy at 3.621 $\mu A{\cdot}cm^{-2}$. After the dynamic potential polarization test, the surface morphology of the alloy is shown in Figure 7. Figure 8a shows the corrosion morphology of the as-cast alloy, which shows that the corrosion form is pitting. Relative to the as-cast alloy, the laser-deposited alloy exhibits shallower pitting pits and shows signs of intergranular corrosion, as depicted in Figure 8b. These results indicate that the laser-deposited alloy exhibits superior corrosion resistance.

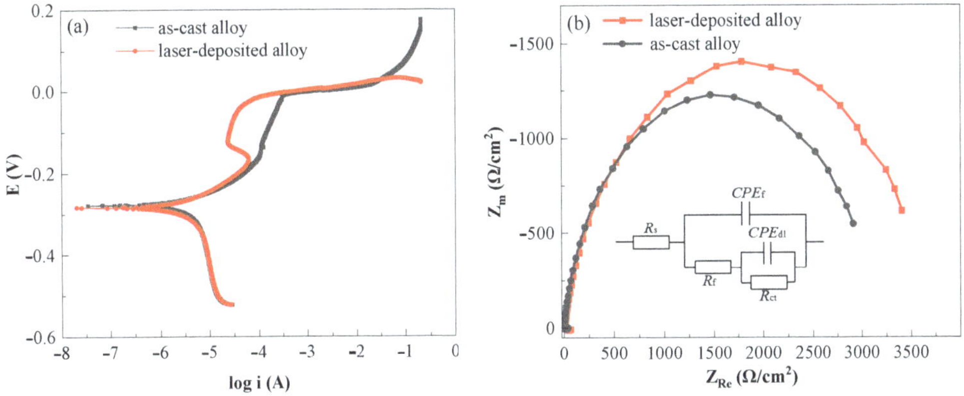

Figure 7. HEA samples in 3.5% NaCl solution showing (**a**) dynamic potential polarization curves; (**b**) Nyquist plots.

Table 2. Fitting parameters of polarization curves for HEA samples in 3.5% NaCl solution.

Sample	E_{corr} (V)	I_{corr} ($\mu A \cdot cm^{-2}$)
As-cast alloy	−0.285	4.062
Laser-deposited alloy	−0.281	3.621

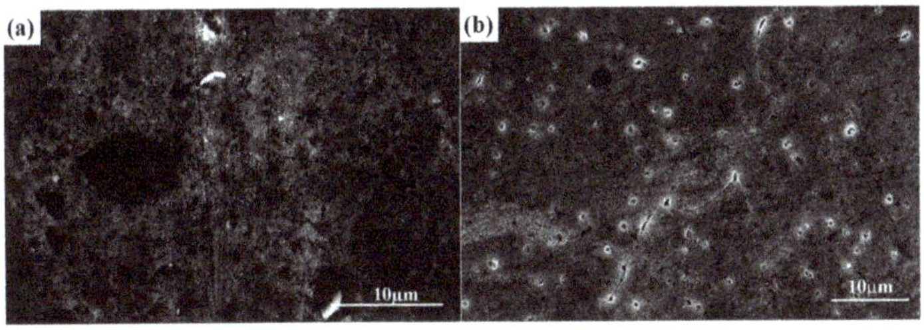

Figure 8. Corrosion morphology of FeCoNiCrAl0.8Cu0.5Si0.5 HEA: (**a**) as-cast alloy; (**b**) laser-deposited alloy.

As demonstrated in the impedance spectra in Figure 6b, the capacitive reactance arc radius of the laser-deposited alloy is larger. The capacitive reactance arc radius is related to the resistance of the corrosion process; a larger radius signifies greater resistance to the reaction and, consequently, a lower corrosion rate [37]. An equivalent circuit with two-time constants is used to extract quantitative information from the EIS, as shown in Table 3. The mathematical expression for an equivalent circuit is as follows [38]:

$$Z = R_s + 1/j\{\omega CPE_f + 1/[R_f + 1/(j\omega CPEf + 1/R_{ct}]\}$$

Table 3. Equivalent circuit-fitting data for FeCoNiCrAl0.8Cu0.5Si0.5 HEA.

Sample	Rs ($\Omega\cdot cm^2$)	$C_{eff\cdot d}$ ($\mu F/cm^2$)	n_d	R_{ct} ($\Omega\cdot cm^2$)	$C_{eff\cdot ct}$ ($\mu F/cm^{-2}$)	n_f	R_{ct} ($\Omega\cdot cm^2$)
As-cast alloy	14.92	27.01	1	341.1	60.4	0.724	2847
Laser-deposited alloy	19.69	12.01	1	372.7	46.68	0.686	3511

Z is the impedance of the circuit; ω is the angular frequency; j is the imaginary unit. The CPE is defined as [38]:

$$CPE = 1/Y_0(j\omega)^n$$

Y_0 is the prefactor of the CPE; n is the exponent of the CPE ($-1 \leq n \leq 1$). When n = 1, −1, and 0 [39], the CPE acts as an ideal capacitor, inductor, and resistor, respectively. A larger n value indicates a smoother passive film, smaller CPE values, and fewer defects in a homogeneously structured passive film [40]. In the equivalent circuit, the CPE is typically used to describe ideal capacitive behavior [41]. The effective capacitance value is independent of the solution conditions and provides a more reliable description of electrochemical performance than the CPE [42]. The mathematical expression for converting CPE to C_{eff} is given as [43]: $C_{eff} = CPE(f_m)^{n-1}$, where f_m is the frequency at which the imaginary part of the impedance reaches its maximum value. The charge transfer resistance and effective capacitance are indicators of an alloy's corrosion resistance. A higher R_{ct} and a lower C_{eff} suggest stronger corrosion resistance. The laser-deposited alloy exhibits the largest R_{ct} and the smallest C_{eff}. The EIS results further demonstrate the superior corrosion resistance of the laser-deposited alloy.

4. Conclusions

FeCoNiCrAl0.8Cu0.5Si0.5 high-entropy alloys (HEAs) were fabricated using both vacuum arc melting and laser deposition techniques to investigate the effects of these distinct processes on the microstructure and properties of the alloys. The following conclusions were drawn:

(1) Phase Structure:

As-cast Alloy: The phase structure consists of BCC1, BCC2, and Cr3Si phases.
Laser-deposited Alloy: The phase structure consists of BCC1 and BCC2 phases.

(2) Microstructure:

As-cast Alloy: The microstructure comprises dendritic and interdendritic structures, with secondary dendritic arms and particle phases formed on the dendrites.

Laser-deposited Alloy: The microstructure displays a dendritic structure without granular interdendritic regions and with interdendritic segregation of Al, Ni, and Cu elements, while Si is uniformly distributed throughout the structure.

(3) Hardness and Wear Resistance:

As-cast Alloy: Demonstrates a hardness of 586 HV0.2 and a wear rate of 3.5×10^{-7} mm^3/Nm, with abrasive wear as the primary wear mechanism.

Laser-deposited Alloy: Exhibits a slightly lower hardness of 557 HV0.2 and a wear rate of 9.0×10^{-7} mm^3/Nm, also with abrasive wear as the primary wear mechanism.

(4) Corrosion Resistance:

The as-cast alloy exhibited an icorr of 4.062 $\mu A\cdot cm^{-2}$, with pitting as the primary form of corrosion.

Laser-deposited Alloy: Showed an icorr of 3.621 $\mu A\cdot cm^{-2}$, with both pitting and intergranular corrosion observed, demonstrating better overall corrosion resistance compared to the as-cast alloy.

These findings highlight the differences in microstructure, mechanical properties, wear resistance, and corrosion resistance between the as-cast and laser-deposited Fe-CoNiCrAl0.8Cu0.5Si0.5 HEAs. Future research should explore the impact of varying laser processing parameters on these properties to optimize the performance of HEAs for specific applications.

5. Expectation

This study investigated FeCoNiCrAl0.8Cu0.5Si0.5 HEAs using specific laser parameters. The focus was on comparing the wear and corrosion resistance of the as-cast and laser-deposited FeCoNiCrAl0.8Cu0.5Si0.5 HEAs. The FeCo-NiCrAl0.8Cu0.5Si0.5 high-entropy alloys demonstrated excellent wear and corrosion resistance in this study, suggesting a wide range of industrial applications. In the aviation sector, these alloys can be used for turbine blades and engine components, as they maintain structural integrity under extreme temperatures and harsh environments, enhancing efficiency and lifespan. In the automotive sector, they are suitable for engine parts and exhaust systems, performing well in high-temperature and corrosive environments, thereby improving fuel efficiency and reducing maintenance costs. In chemical processing equipment, these alloys' superior corrosion resistance makes them ideal for reactors and pipelines, reducing downtime and repair costs. For marine applications, their enhanced corrosion resistance offers longer service life and lower maintenance for shipbuilding and offshore structures. In the energy sector, these alloys can improve the efficiency and durability of power generation equipment such as heat exchangers and boilers. Based on the findings presented, future research should consider the following aspects:

(1) Investigating the influence of varying laser processing parameters on the microstructure and properties of the alloy to gain a deeper understanding of how these parameters can be optimized for improved material performance.
(2) Expanding the study to explore the effect of different alloy compositions on wear and corrosion resistance to better tailor the material properties for specific industrial applications.

Author Contributions: Conceptualization, X.J.; methodology, X.J.; software, X.J.; validation, X.J.; formal analysis, X.J.; investigation, X.J.; resources, X.J.; data curation, X.J.; writing—original draft preparation, X.J.; writing—review and editing, X.J.; visualization, H.L.; supervision, L.Z.; project administration, X.J.; funding acquisition, H.L. All authors have read and agreed to the published version of the manuscript.

Funding: This study was supported by the High-level Talents Research Project of West Anhui University (Grant Nos. WGKQ2022058) and the Fundamental Research Funds for the Central Universities of China (Grant No.PA2023GDSK0057).

Institutional Review Board Statement: Not applicable.

Informed Consent Statement: Not applicable.

Data Availability Statement: Data are contained within the article.

Conflicts of Interest: The authors declare no conflicts of interest.

erences

Zhu, H.; Zhu, Z.; Hao, J.; Sun, S.; Lu, S.; Wang, C.; Ma, P.; Dong, W.; Du, M. High-entropy alloy stabilized active Ir for highly efficient acidic oxygen evolution. *Chem. Eng. J.* **2022**, *431*, 133251. [CrossRef]
Cui, D.; Zhang, Y.; Liu, L.; Li, Y.; Wang, L.; Wang, Z.; Li, J.; Wang, J.; He, F. Oxygen-assisted spinodal structure achieves 1.5 GPa yield strength in a ductile refractory high-entropy alloy. *J. Mater. Sci. Technol.* **2023**, *157*, 11–20. [CrossRef]
Roche, V.; Champion, Y.; Bataev, I.A.; Jorge, A.M. Passive film formation on the new biocompatible non-equiatomic$Ti_{21}Nb_{24}Mo_{23}Hf_{17}Ta_{15}$ high entropy alloy before and after resting in simulated body fluid. *Corros. Sci.* **2022**, *207*, 110607. [CrossRef]
Cui, K.X.; Qiao, J.W.; Liaw, P.K.; Zhang, Y. Data driving design of high-entropy alloys for lightweight and dynamic applications. *Sci. China-Phys. Mech. Astron.* **2024**, *67*, 227101. [CrossRef]

5. Ghosh, P.S.; Ali, K.; Arya, A. Efficient screening of single phase forming low-activation high entropy alloys. *J. Alloys Compd.* **20** *978*, 173172. [CrossRef]
6. Wang, B.Y.; Yang, C.; Shu, D.; Sun, B.D. A Review of Irradiation-Tolerant Refractory High-Entropy Alloys. *Metals* **2024**, *14*, [CrossRef]
7. Zhang, C.; Zhao, Z.; Lin, X.; Wang, S.; Wang, J.; Li, Y.; Li, Y.; Zhang, Y.; Zhao, H. Molybdenum-14Rhenium Alloy—The M Promising Candidate for High-Temperature Semiconductor Substrate Materials. *J. Alloys Compd.* **2024**, *991*, 174391. [CrossRe
8. Dong, D.; Jiang, W.T.; Wang, X.H.; Ma, T.F.; Zhu, D.D.; Wang, Y.; Huo, J.T. Enhanced mechanical properties of high pressu solidified $CoCrFeNiMo_{0.3}$ high entropy alloy via nano-precipitated phase. *Intermetallics* **2024**, *166*, 108192. [CrossRef]
9. Mazitov, A.; Springer, M.A.; Lopanitsyna, N.; Fraux, G.; De, S.D.; Ceriotti, M. Surface segregation in high-entropy alloys fr alchemical machine learning. *J. Phys. Mater.* **2024**, *7*, 025007. [CrossRef]
10. Nagase, T.; Rack, P.D.; Noh, J.H.; Egami, T. *In-situ* TEM observation of structural changes in nano-crystalline CoCrCuFeNi mu component high-entropy alloy (HEA) under fast electron irradiation by high voltage electron microscopy (HVEM). *Intermetall* **2015**, *59*, 32–42. [CrossRef]
11. Cheng, J.B.; Liang, X.B.; Xu, B.S. Effect of Nb addition on the structure and mechanical behaviors of CoCrCuFeNi high-entro alloy coatings. *Surf. Coat. Technol.* **2014**, *240*, 184–190. [CrossRef]
12. Zhang, Y.; Ma, S.G.; Qiao, J.W. Morphology Transition from Dendrites to Equiaxed Grains for AlCoCrFeNi High-Entropy All by Copper Mold Casting and Bridgman Solidification. *Metall. Mater. Trans. A-Phys. Metall. Mater. Sci.* **2012**, *43A*, 2625–26 [CrossRef]
13. Ma, S.G.; Zhang, Y. Effect of Nb addition on the microstructure and properties of AlCoCrFeNi high-entropy alloy. *Mater. Sci. E A-Struct. Mater. Prop. Microstruct. Process.* **2012**, *532*, 480–486. [CrossRef]
14. Qiao, J.W.; Ma, S.G.; Huang, E.W.; Chuang, C.P.; Liaw, P.K.; Zhang, Y. Microstructural characteristics and mechanical behaviors AlCoCrFeNi high-entropy alloys at ambient and cryogenic temperatures. In *Nano-Scale And Amourphous Materials;* Trans Te Publications Ltd.: Bach, Switzerland, 2011; pp. 419–425.
15. Lu, Y.P.; Dong, Y.; Guo, S.; Jiang, L.; Kang, H.J.; Wang, T.M.; Wen, B.; Wang, Z.J.; Jie, J.C.; Cao, Z.Q.; et al. A Promising New Cl of High-Temperature Alloys: Eutectic High-Entropy Alloys. *Sci. Rep.* **2014**, *4*, 6200. [CrossRef] [PubMed]
16. Xiong, T.; Zheng, S.J.; Pang, J.Y.; Ma, X.L. High-strength and high-ductility $AlCoCrFeNi_{2.1}$ eutectic high-entropy alloy achie via precipitation strengthening in a heterogeneous structure. *Scr. Mater.* **2020**, *186*, 336–340. [CrossRef]
17. Nassar, A.; Mullis, A.; Cochrane, R.; Aslam, Z.; Micklethwaite, S.; Cao, L. Rapid solidification of $AlCoCrFeNi_{2.1}$ High-entro Alloy. *J. Alloys Compd.* **2022**, *900*, 163350. [CrossRef]
18. Shi, P.J.; Ren, W.L.; Zheng, T.X.; Ren, Z.M.; Hou, X.L.; Peng, J.C.; Hu, P.F.; Gao, Y.F.; Zhong, Y.B.; Liaw, P.K. Enhanced streng ductility synergy in ultrafine-grained eutectic high-entropy alloys by inheriting microstructural lamellae. *Nat. Commun.* **20** *10*, 489. [CrossRef]
19. Wang, J.T.; Long, Z.P.; Jiang, P.F.; Fautrelle, Y.; Li, X. Microstructure, Crystallographic Orientation and Mechanical Prope in $AlCoCrFeNi_{2.1}$ Eutectic High-Entropy Alloy Under Magnetic Field-Assisted Directional Solidification. *Metall. Mater. Tra A-Phys. Metall. Mater. Sci.* **2020**, *51*, 3504–3517. [CrossRef]
20. Shi, P.J.; Zhong, Y.; Li, Y.; Ren, W.L.; Zheng, T.X.; Shen, Z.; Yang, B.; Peng, J.C.; Hu, P.F.; Zhang, Y.; et al. Multistage work hardeni assisted by multi-type twinning in ultrafine-grained heterostructural eutectic high-entropy alloys. *Mater. Today* **2020**, *41*, 62– [CrossRef]
21. Wani, I.S.; Bhattacharjee, T.; Sheikh, S.; Bhattacharjee, P.P.; Guo, S.; Tsuji, N. Tailoring nanostructures and mechanical propert of $AlCoCrFeNi_{2.1}$ eutectic high entropy alloy using thermo-mechanical processing. *Mater. Sci. Eng. A-Struct. Mater. P Microstruct. Process.* **2016**, *675*, 99–109. [CrossRef]
22. Hsu, C.Y.; Sheu, T.S.; Yeh, J.W.; Chen, S.K. Effect of iron content on wear behavior of $AlCoCrFe_xMo_{0.5}Ni$ high-entropy allo *Wear* **2010**, *268*, 653–659. [CrossRef]
23. Kudrnovsky, J.; Drchal, V.; Máca, F.; Turek, I.; Khmelevskyi, S. Electron transport in high-entropy alloys: $Al_xCrFeCoNi$ as a ca study. *Phys. Rev. B* **2019**, *100*, 014441. [CrossRef]
24. Torralba, J.M.; Campos, M. High Entropy Alloys Manufactured by Additive Manufacturing. *Metals* **2020**, *10*, 639. [CrossRef]
25. Wei, S.M.; Ma, P.; Ji, P.C.; Ma, Y.C.; Wang, C.; Zhao, J.; Yu, Z.S. Research progress in high entropy alloys by additive manufacturi *Cailiao Gongcheng-J. Mater. Eng.* **2021**, *49*, 1–17. [CrossRef]
26. Buranich, V.; Rogoz, V.; Postolnyi, B.; Pogrebnjak, A. Predicting the Properties of the Refractory High-Entropy Alloys for Additi Manufacturing-Based Fabrication and Mechatronic Applications. In Proceedings of the 2020 IEEE 10th International Conferer On Nanomaterials: Applications & Properties (Nap-2020), Sumy, Ukraine, 9–13 November 2020.
27. Guo, Y.N.; Su, H.J.; Zhou, H.T.; Shen, Z.L.; Liu, Y.; Zhang, J.; Liu, L.; Fu, H.Z. Unique strength-ductility balance of AlCoCrFeN eutectic high entropy alloy with ultra-fine duplex microstructure prepared by selective laser melting. *J. Mater. Sci. Technol.* **20** *111*, 298–306. [CrossRef]
28. He, L.; Wu, S.W.; Dong, A.P.; Tang, H.B.; Du, D.F.; Zhu, G.L.; Sun, B.D.; Yan, W.T. Selective laser melting of dense and crack-f $AlCoCrFeNi_{2.1}$ eutectic high entropy alloy: Synergizing strength and ductility. *J. Mater. Sci. Technol.* **2022**, *117*, 133–145. [CrossR
29. Ren, J.; Zhang, Y.; Zhao, D.X.; Chen, Y.; Guan, S.; Liu, Y.F.; Liu, L.; Peng, S.Y.; Kong, F.Y.; Poplawsky, J.D.; et al. Strong yet duc nanolamellar high-entropy alloys by additive manufacturing. *Nature* **2022**, *608*, 62–68. [CrossRef] [PubMed]

Chen, X.S.; Kong, J.; Li, J.L.; Feng, S.; Li, H.; Wang, Q.P.; Liang, Y.Z.; Dong, K.W.; Yang, Y. High-strength $AlCoCrFeNi_2$. *1 eutectic high entropy alloy with ultrafine lamella structure via additive manufacturing. Mater. Sci. Eng. A-Struct. Mater. Prop. Microstruct. Process.* **2022**, *854*, 143816. [CrossRef]
Li, Y.; Shi, Y.; Wang, H.; Zhou, B.; Li, D.; Lin, H.; Wang, J. Effect of Y on Microstructure and Properties of Al0.8FeCrCoNiCu0.5 High Entropy Alloy Coating on 5083 Aluminum by Laser Cladding. *Lubricants* **2023**, *11*, 50. [CrossRef]
Dias, D.A.; Lima, F.W.S.; Plascak, J.A. Generalized Gibbs Phase Rule and Multicriticality Applied to Magnetic Systems. *Entropy* **2022**, *24*, 63. [CrossRef]
Ge, H.H.; Li, J.; Han, X.J.; Xia, M.X.; Li, J.G. Dendritic model for macrosegregation prediction of large scale castings. *J. Mater. Process. Technol.* **2016**, *227*, 308–317. [CrossRef]
Wang, L.M.; Chen, C.C.; Yeh, J.W.; Ke, S.T. The microstructure and strengthening mechanism of thermal spray coating $Ni_xCo_{0.6}Fe_{0.2}Cr_ySi_zAlTi_{0.2}$ high-entropy alloys. *Mater. Chem. Phys.* **2011**, *126*, 880–885. [CrossRef]
Li, L.L.; Kamachali, R.D.; Li, Z.M.; Zhang, Z.F. Grain boundary energy effect on grain boundary segregation in an equiatomic high-entropy alloy. *Phys. Rev. Mater.* **2020**, *4*, 053603. [CrossRef]
Chen, Y.; Li, X.; Liu, J.; Zhang, Y.; Chen, X. EFFECT OF LASER SURFACE REMELTING OF 304 STAINLESS STEEL. *J. Appl. Mech. Tech. Phys.* **2023**, *64*, 491–498. [CrossRef]
Li, Y.Z.; Shi, Y.; Lina, H.; Ji, X.H. Corrosion Behavior of As-Cladding $Al_{0.8}CrFeCoNiCu_{0.5}Six$ High Entropy Alloys in 3.5% NaCl Solution. *Mater. Res.-Ibero-Am. J. Mater.* **2024**, *27*, e20230221. [CrossRef]
Ji, P.F.; Li, B.; Chen, B.H.; Wang, F.; Ma, W.; Zhang, X.Y.; Ma, M.Z.; Liu, R.P. Effect of Nb addition on the stability and biological corrosion resistance of Ti-Zr alloy passivation films. *Corros. Sci.* **2020**, *170*, 108696. [CrossRef]
Wasekar, N.P.; Hebalkar, N.; Jyothirmayi, A.; Lavakumar, B.; Ramakrishna, M.; Sundararajan, G. Influence of pulse parameters on the mechanical properties and electrochemical corrosion behavior of electrodeposited Ni-W alloy coatings with high tungsten content. *Corros. Sci.* **2020**, *165*, 108409. [CrossRef]
Quintero, D.; Galvis, O.; Calderón, J.A.; Castaño, J.G.; Echeverría, F. Effect of electrochemical parameters on the formation of anodic films on commercially pure titanium by plasma electrolytic oxidation. *Surf. Coat. Technol.* **2014**, *258*, 1223–1231. [CrossRef]
Jayaraj, J.; Shankar, A.R.; Mudali, U.K. Electrochemical and passive characterization of a beta type $Ti_{45}Zr_{38}Al_{17}$ cast rod in nitric acid medium. *Electrochim. Acta* **2012**, *85*, 210–219. [CrossRef]
Potucek, R.K.; Rateick, R.G.; Birss, V.I. Impedance characterization of anodic barrier Al oxide film beneath porous oxide layer. *J. Electrochem. Soc.* **2006**, *153*, B304–B310. [CrossRef]
Hsu, C.H.; Mansfeld, F. Technical note: Concerning the conversion of the constant phase element parameter Y_0 into a capacitance. *Corrosion* **2001**, *57*, 747–748. [CrossRef]

coatings

MDPI

Article

In Situ Thermal Interactions of Cu-Based Anti-Corrosion Coatings on Steel Implemented by Surface Alloying

Huda Hanif Khan [1,†], Tong Wang [1,†], Lihong Su [1], Huijun Li [1], Qiang Zhu [1], Ana Yang [2], Zigang Li [2], Wei Wang [2,*] and Hongtao Zhu [1,*]

1 School of Mechanical Materials Mechatronics and Biomedical Engineering, University of Wollongong, Wollongong, NSW 2522, Australia; hhk991@uowmail.edu.au (H.H.K.); tw874@uowmail.edu.au (T.W.); lihongsu@uow.edu.au (L.S.); huijun@uow.edu.au (H.L.); qzhu@uow.edu.au (Q.Z.)

2 Baosteel Central Research Institute, Baowu Steel Group Corporation, Shanghai 201900, China; yangan@baosteel.com (A.Y.); zigangli@baosteel.com (Z.L.)

* Correspondence: weiwang@baosteel.com (W.W.); hongtao@uow.edu.au (H.Z.)

† These authors contributed equally to this work.

Abstract: Incorporating expensive alloying elements into bulk steel for corrosion protection is undesirable, considering that only the surfaces are exposed to aggressive environments. Therefore, this work focused on developing and optimizing a new surface functioning technology through in situ observation of thermal interactions between the metallic powders at elevated temperatures. The study revealed that the Cu-Ni powder mixture, with 12.5 wt% Ni, began to melt at 1099.5 °C and was fully melted at 1175 °C, significantly different from the Cu-Ni solid solution and bulk Cu or Ni. As a result of high-temperature reactions, copper penetration of up to 35 μm for pure copper and 55 μm for copper-chromium composite coatings occurred due to liquid metal corrosion. In contrast, the copper-nickel composite coating exhibited a cupronickel solution microstructure with FeNi dendrites and a nickel-rich transition layer. This cupronickel coating, with a chemical composition of 89.3 wt% Cu, 6.2 wt% Ni, and 4.5 wt% Fe, demonstrated uniform thickness, superior surface morphology, and continuous coverage on the steel substrate. Furthermore, the Ni-rich transition layer played a vital role in preventing copper penetration along the grain boundary of the steel matrix while forming a chemical binding between the coating and the substrate. The practicality of the coating was further confirmed through the hot-rolling procedure and subsequent electrochemical corrosion tests, which resulted in a 44% improvement in corrosion resistance.

Keywords: surface alloying; anti-corrosion coating; cupronickel

1. Introduction

Corrosion poses a significant threat to the structural integrity of marine and offshore assets, which results in surface degradation, mechanical failure, environmental pollution, and loss of life. Steel, being a primary material used in various industries such as transportation, oil and gas, petrochemical, construction, and marine engineering, has been subject to numerous corrosion prevention methods throughout history [1], such as corrosion inhibitors, alloying, sacrificial anode, and protective coatings. Protective coatings, in particular, have proven to be effective in extending the service life of steel structures [2,3]. Examples of protective coatings include paints, organic/inorganic coatings, diffusion coatings, thermal spray metallic coatings, and galvanization [4–7]. Alloying, another essential methodology, involves adding alloying elements such as Cu, Cr, Ni, P, Si, and Mn to bulk steel. These elements help form a protective passivation layer, known as patina, on the surface when exposed to a corrosive environment, thus protecting the material underneath [8]. Both protective coatings and alloying have their advantages and limitations. Applying protective coatings is time-consuming and expensive, requiring complex procedures for surface preparation and application [9,10]. Alloying, on the other hand, involves adding expensive elements to the bulk material when only the surface is exposed to harsh environments. Moreover, alloying could alter the material's mechanical properties of the bulk material, which is not always desirable.

Cu, Cr, and Ni are widely used alloying elements to enhance the corrosion resistance of steels. These elements facilitate the formation of an adhesive oxide layer when exposed to aggressive environments, thereby improving the product's service life [11]. Due to its low cost, copper is commonly introduced as a base alloying element. However, adding Cu can lead to hot shortness, a complex high-temperature chemical reaction process. The relatively low melting Cu-enriched phase liquefies at high temperatures and penetrates along the austenite grain boundaries of the steel matrix, resulting in cracking during processes like hot-rolling [12]. Ni plays a crucial role in preventing hot shortness by being noble compared to Fe, exhibiting less tendency to oxidize when oxide scales form on the steel surface. Furthermore, Ni and Cu are completely soluble in each other, increasing the melting point of the Cu-enriched liquid phase and restricting it from penetrating grain boundaries. Akamatsu et al. [13] found that the liquid phase decreases as the percentage of Ni increases, and no more liquid phase is observed when the percentage of Ni is equal to or greater than that of Cu. Yamamoto et al. [14] developed a new surface melting technology whereby the surface layer of the slab at the back of the continuous caster is uniformly melted utilizing plasma or electromagnetic heating. Ni, as a wire, is added as an alloying agent to the melted surface layer to prevent Cu in the surface layer from promoting hot shortness at subsequent rolling. The surface melting technology proves that Ni and Cu can be introduced on steel surfaces for anti-corrosion without incurring adverse hot shortness. Cupronickels, alloys of Cu and Ni, are known for their high resistance to corrosion and biofouling properties [15], which are widely used in marine engineering, coinage, and cryogenic applications [16]. Tolulope Loto [17] found that Cu-Ni 90/10 exhibited superior corrosion resistance at a high NaCl concentration solution, i.e., the corrosion rate of Cu-Ni 90/10 was just 0.74 mm/yr compared to the 2.68 mm/yr value of Cu–Zn 70/30 alloy at 4.5% NaCl concentration. The addition of Cr in steel is beneficial [18]. S. A. Park et al. [19] conducted electrochemical corrosion tests on low-alloy steel containing 0.1%, 0.3%, and 0.5% Cr after immersion in a chloride-containing environment for seven days and demonstrated that the addition of up to 0.3% Cr positively affected corrosion behavior, whereas 0.5% Cr promoted localized corrosion.

Surface alloying is an exciting technique to alter steel's surface properties without compromising the properties of the bulk material. It has been implemented for several different purposes, such as enhancing surface hardness, scratch resistance, corrosion resistance, etc. Carburizing, nitriding, boriding, chromizing, and aluminizing are examples of conventional surface alloying methods. Typically, this process involves heating the material at high temperatures and introducing the desired alloying element into the atmosphere in a gaseous or plasma state or through a sand bath. However, the material needs to be heated for several hours in the atmosphere to achieve the desired diffusion thickness [20]. It is crucial to select the appropriate temperatures and understand the thermal interactions between the alloying element particles and the steel surface at those temperatures when designing a surface alloying technique [21–23]. For example, the fully melted Cu-based coating could increase the risk of liquid metal embrittlement, leading to crack formation or complete failure during service [24–26]. In contrast, insufficient heating would not result in a solid solution for successful surface alloying [23].

Thermal analysis software such as Thermo-Calc can predict alloys' phase transformations and thermal behaviors based on their compositions under equilibrium conditions. However, these calculations rely on equilibrium conditions and are based on pre-alloyed materials that may or may not be achieved during the rapid heating and cooling of metallic powders. A high-temperature confocal laser scanning microscope has been extensively used by various researchers to observe in situ the thermal behaviors of alloys during heating or cooling processes [27–30]. It can capture real-time images of the specimen as it undergoes solidification or transforms into a liquid phase. It has been successfully employed to study phase transformations, slag formation, dissolution of elements/compounds, and other helpful information [31,32]. Raman spectroscopy has also been widely used for in situ and operando exploration of material properties, real-time monitoring of high-temperature reactions, and understanding of high-temperature heat treatments. Böhme et al. [33] used a heating stage coupled with a Raman microscope to investigate the formation and stability properties of hydroxylapatite in situ. Molten salt chemistry has been explored using ultraviolet-visible (UV-vis) absorbance and Raman spectroscopy [34]. More recently, Zhang et al. [35,36] developed a customized in situ high-temperature fiber-optic Raman probe to reveal the structure of glass and slag and investigate in situ molten flux for continuous casting in steel manufacturing at temperatures up to 1400 °C.

Current research focuses on developing and optimizing a new surface alloying technology that can be implemented on commonly hot-rolled carbon steels by integrating Cu-based metallic compounds and steel matrix to form a corrosion-resistant surface coating. Unlike adding alloying elements to the bulk material, this technique solely alloys the surface, creating a protective coating without altering the properties of the bulk material. Surface alloying of steel at a high temperature presents a promising solution to improve the corrosion resistance of carbon steel at a low manufacturing cost.

The proposed approach combines and modifies surface diffusive coatings and thermal spray coating techniques to create a more cost-effective method that can be integrated into steel strip and plate manufacturing, eliminating the need for additional processing costs [37]. In conventional diffusion coating, the substrate surface gets enriched by the diffusing elements. However, no physical coating is developed, and the surface becomes an alloy of the substrate and the diffusing elements [38,39]. On the other hand, thermal spray coatings involve spraying hot coating materials onto the cold substrate surface, creating a layer on top, but adhesiveness is not as good as what could be achieved with the diffusion technique. Furthermore, the significant temperature difference between hot sprayed material and cold substrate can lead to the formation of cracks within the coating and coating-substrate interface, compromising the corrosion resistance of the coating [40–42]. Therefore, it is vital to combine and modify these methods by heating the alloying powders on the steel substrate to form a protective coating with a diffused interface. Moreover, the traditional surface alloying techniques are primarily designed for single-element coatings, whereas the proposed method can accommodate a combination of a few elements within a single coating process. For example, a cupronickel compound coating containing two metallic elements, Cu and Ni, is developed. Theoretical results from Thermo-Calc software 2024a with Database 'TCFE9' were used to design the experimental procedure and heating profiles. Based on these heating profiles, tests were conducted, and in situ observations were recorded using high-temperature confocal scanning laser microscopy.

2. Materials and Methodology

2.1. Materials

Coating materials were chosen by considering the effect of alloying elements on the corrosion properties of the steel, with particular attention to the synergistic impact of multi-elements [43]. Current work developed a new surface mitigation technology that integrates Cu, Ni, and Cr metallic powders with a steel matrix to form a robustly bonded, corrosion-resistant layer onto steel substrates. Cu is a base anti-corrosion element due to its cost-effective and appreciated melting point of 1083 °C, closely aligning with the working temperature in the hot rolling of carbon steel plates or strips [44,45]. The addition of Ni not only enhances the solubility/diffusion of Cu in the γ phase but also increases the melting point of the Cu-rich compound, which prevents the adverse impact of hot shortness [46]. The inclusion of Cr helps improve the corrosion resistance; however, the high affinity of Cr could make it sensitive to high-temperature oxidation. It is essential to report that both Cr and Ni have higher melting points, 1907 °C and 1455 °C, respectively, which can help to maintain a solid-liquid state at working temperature instead of a fully molten liquid. Since the alloying elements are only applied to the surface of the steel instead of the bulk of the steel, this approach offers significant cost savings.

Q235 steel (99.13% Fe, 0.1637% C, 0.67 Mn, 0.022 Si, and traces of sulfur) was introduced as the steel substrate for the current investigation. Q235 steel and its ASTM equivalent A36 are prevalent in structural applications, such as bridges and boiler support structures, etc., due to its superior plasticity, toughness, weldability, cold bending performance, and adequate strength, along with low price compared to other steel grades and alloyed steels.

In this research, we utilized Cu powder (particle size of 45 μm, purity of 99%, and density of 3–3.8 g/cm^3), Cr powder (particle size of 5 μm, purity of 99%), and Ni powder (particle size of 1 μm, purity of 99%). Three distinct coating powder combinations, including (1) pure Cu, (2) a mixture of 87.5 wt% Cu and 12.5 wt% Ni, and (3) a mixture of 90 wt% Cu and 10 wt% Cr, were developed to explore the proposed surface alloying techniques. Each component was individually weighed using a physical scale and then mixed thoroughly together.

The ratio of Cu and Ni mixture was selected based on cupronickel 90/10 alloy, known for its exceptional corrosion-resistive properties. Furthermore, a marginally increased proportion of Ni to 12.5 wt% was chosen due to Ni's solubility in Fe at high temperatures, potentially forming a barrier that prevents Cu from penetrating deeper into the substrate's grain boundaries. The Cu–Cr weight ratio was maintained at 9:1, a slight reduction of Cr compared to Ni in the Cu-Ni mixture, taking into account the higher melting point of Cr and its insolubility in Cu.

2.2. Experimental Setup for In Situ Observation and Designing Thermal Profiles

The in situ and real-time thermal interactions of the designed metallic coatings were investigated with a high-temperature confocal laser scanning microscope (HTCLSM), which simulates the operational conditions of an industrial reheating furnace in a hot strip mill. The HTCLSM, manufactured by Lasertec Corporation, Yokohama, Japan, features a vacuum chamber-type heating furnace that allows observation under various conditions, such as in a vacuum, with inert gas, or in a reducing atmosphere. The confocal optics enable the observation and video recording of the melting and solidification process without interference from the high-intensity light emitted at elevated temperatures. The observation and recording of the melting and solidification process of Cu-coated samples were correlated with the actual temperature and thermal history/profiles employed during heating and cooling. This correlation helps in understanding phase transitions and microstructural evolution influenced by specific heating and cooling profiles and rates.

Round, thin Q235 specimens of 5.5 mm in diameter and 2.2 mm in thickness were sectioned from the steel plate and then polished using a Struers Tegramin grinding and polishing machine (Copenhagen, Denmark) for in situ high-temperature observation.

Cu-Ni alloy is recognized as an isomorphous system known for its ability to form solid solutions over a range of temperatures [47]. As shown in Figure 1, we used Thermo-Calc software to calculate the Cu-Ni binary phase diagram. The results indicate that the solidus and liquidus temperatures for the designed Cu-Ni coating with a 12.5 wt% Ni are 1137.5 °C and 1162.5 °C, respectively. In Figure 1, α_1 represents a Cu solid solution with an fcc structure and α_2 denotes a Ni fcc-solid solution, which shows a miscibility gap in the Cu-Ni phase diagram at lower temperatures. Hence, a temperature above the liquidus temperature, 1250 °C, was selected to investigate the melting and solidification process through HTCLSM. This temperature is commonly employed in actual heating and reheating furnaces for hot steel manufacturing. The same heating temperature was chosen for the Cu–Cr coating as well. However, for the pure Cu powder coating, a slightly low temperature of 1200 °C was selected, considering the low melting point of Cu. In contrast, the Cu–Cr binary system exhibits very limited solubility between the two elements, with nearly no solubility up to 1076 °C.

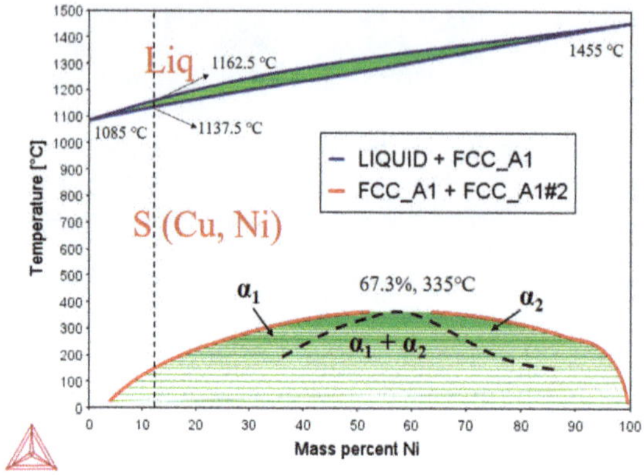

Figure 1. Cu-Ni binary phase diagram calculated by Thermo-Calc software.

Figure 2 presents our designed thermal profiles of three coatings used in HTCLSM experiments. In the pure Cu experiment, as shown in Figure 2a, the Q235 substrate and coating powder were heated to 800 °C at a rate of 100 °C/min, which is below Cu's melting point of 1085 °C. Following this, the temperature was increased to 1200 °C at a reduced rate of 50 °C/min. Then, the temperature was maintained at 1200 °C for 10 min before cooling down to 800 °C at a rate of 10 °C/min and followed by rapid cooling to 200 °C at a rate of 100 °C/min. For the samples covered with Cu-Ni and Cu–Cr powder composites, as depicted in Figure 2b, the samples were heated to 1250 °C at a rate of 100 °C/min and then held at this temperature for 10 min. After that, the samples were cooled to 1050 °C with a 20 °C/min cooling rate and then further reduced to 200 °C at a cooling rate of 100 °C/min. The HTCLSM furnace chamber was filled with argon gas at a flow rate of 0.4 LPM to protect the samples from severe oxidation during the testing process.

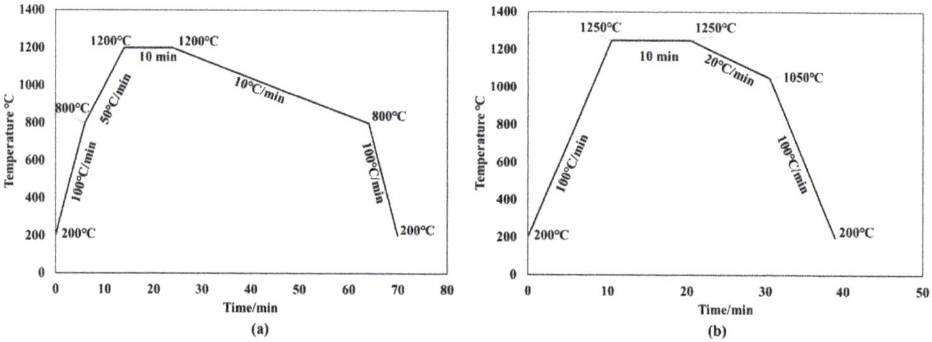

Figure 2. Thermal setup of (**a**) pure Cu, (**b**) Cu-Ni, and Cu–Cr coatings for HTCLSM experiments.

2.3. Coating Performance Characterization

Following the in situ observation, the top surface and cross-section of the samples along the radial plane were exanimated using optical microscopy (OM), scanning electron microscopy (SEM), and energy dispersive spectrometry (EDS) in terms of surface morphology, coating thickness, microstructure, and the bonding and diffusion behaviors between the coating layer and Q235 substrate.

3. Results and Discussion

3.1. In Situ Phase Transformation of Coatings

The melting and solidification behavior of Cu, Cu-Ni, and Cu–Cr composite coatings on Q235 steel substrates were influenced by the microstructure, phase composition, thermal properties, and interfacial interactions between the composites and the steel substrate.

Figure 3 displays the melting and solidification process of the pure Cu-coated samples, subject to the thermal history detailed in Figure 2a. It can be found that Cu powder was uniformly dispersed on the top surface of the Q235 substrate, as evidenced by an in situ image at 641 °C prior to melting. The in situ observation revealed that the Cu powder started melting at 1050 °C, slightly below the melting point of the bulk Cu at 1083 °C. Figure 3b illustrates that a few liquid Cu pools had formed at 1182.3 °C. As shown in Figure 3c, the Cu powder has fully melted after holding at 1200 °C for ten minutes, and some isolated oxides in both liquid and solid states were spotted floating on the liquid Cu pool. Figure 3d presents the solidification process of liquid oxides. As the temperature decreased, the crystallization started around the marked oxides and then proceeded in various directions on the surface of the liquid Cu pool. This phenomenon is further evidenced by the detailed inserted images from 1164.8 °C to 1127.0 °C in Figure 3d.

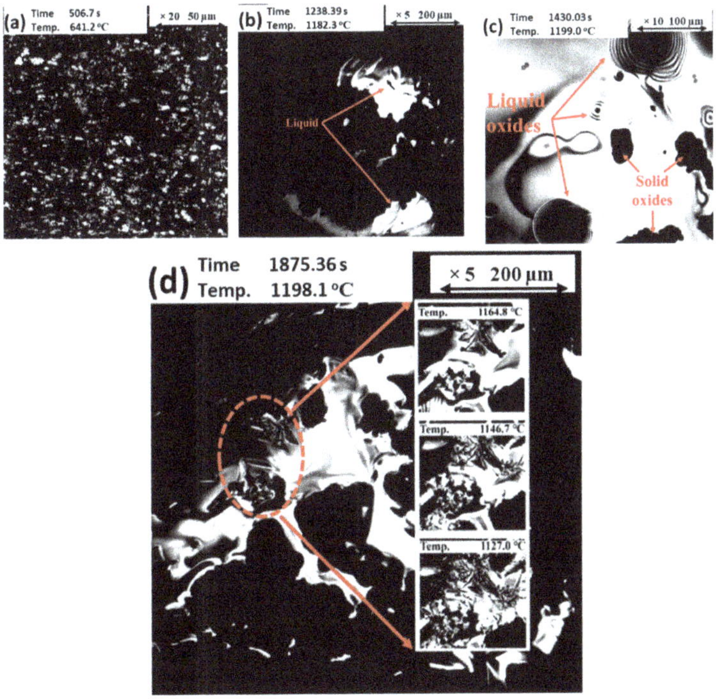

Figure 3. Melting and solidification process of pure Cu-coated sample. (**a**) Cu powder on the surface, (**b**) Melting started, (**c**) Fully melted, (**d**) Solidification.

Figure 4 shows the solidification process of the Cu–Cr coating. The melting point of bulk Cr is around 1907 °C, which is significantly higher than the 1493 °C of bulk Q235. As highlighted within the red dot circle in Figure 4a, a considerable number of solid particles still remained within the liquid phase, even after holding at 1250 °C for ten minutes. These solid particles were later identified as unmelted Cr powders during subsequent material characterization. This phenomenon is substantiated by the extremely low solubility of Cr in Cu. In addition to the unmelted Cr particles, the solid oxide clusters were observed floating on the liquid surface. The solidification process at 960 °C can be found in Figure 4b. It can be seen that most of the coating remained in the liquid phase, which is assumed to be liquid Cu. It proves that the addition of Cr to the Cu powder has slowed down the solidification of the composite coating. As seen in Figure 4c, most of the liquid had transformed into solids, with some residual liquid nestled between solids at 950 °C.

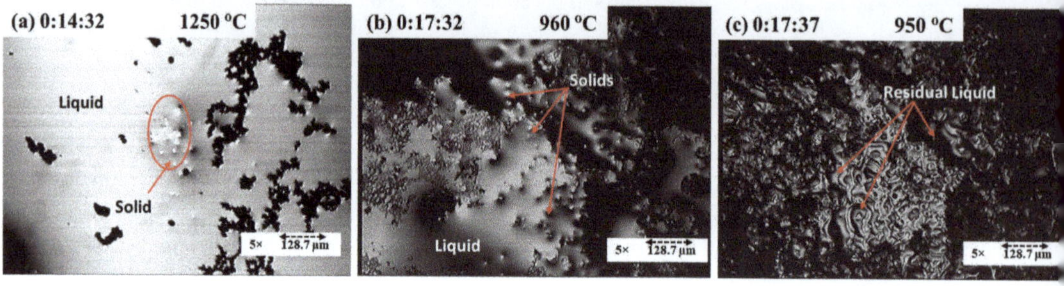

Figure 4. Solidification process of Cu–Cr coating. (**a**) Liquid pool with undissolved solid particles, (**b**) a combination of the solid and liquid phases, (**c**) solidified coating with the lesser residual liquid left.

As per the Cu-Ni phase diagram illustrated in Figure 1, the Cu-Ni solid solution with a 12 %wt Ni should initiate melting at 1137.5 °C. However, the in situ observation in Figure 5a shows that the Cu-Ni powder mixture began to melt at 1099.5 °C, which is far below 1137.5 °C for the Cu-Ni solid solution and marginally higher than 1085 °C for the bulk Cu. This discrepancy provides evidence that the solid-liquid phase transformation of the powder mixture is significantly different from that of the bulk alloy with the same chemical composite, which must be considered in the design of surface alloying. At 1156.7 °C, the melting process became faster. A combination of solid and liquid phases was observed in Figure 5b, which swiftly converted into a liquid pool with minimal solid particles remaining at 1163.6 °C, as illustrated in Figure 5c. Although bulk Ni only started to melt at 1453 °C, the Cu-Ni powder mixture was fully melted at 1175 °C, as shown in Figure 5d. It is vital to report that the melting points observed during this process are higher than that of bulk Cu but significantly lower than that of bulk Ni. The presence of Ni slows down the melting of Cu. Due to Ni's solubility in liquid Cu, the melting point of the Cu-Ni coating is significantly lower than that of bulk Ni. The properties of the substrate also affect the coating's melting and solidification behaviors. Zhao et al. observed that for the same coating material, the coating properties and microstructure varied for aluminum and steel substrates, while the coating parameters were identical [48].

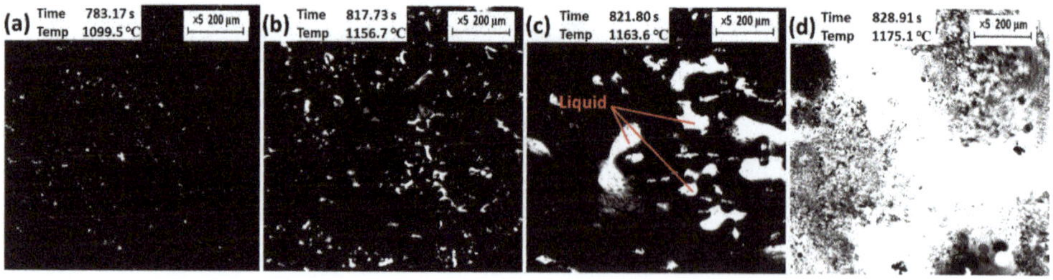

Figure 5. Melting process of Cu-Ni powders. (**a**) Metallic powders started to melt, (**b**) Solid powders with some molten liquid, (**c**) Liquid pool started to form with fewer solid powders left, (**d**) fully melt.

The solidification process of Cu-Ni composite coating on the Q235 substrate is shown in Figure 6. As seen in Figure 6a, after the complete melting of the Cu-Ni powders, some solid and liquid oxides were observed floating on the liquid Cu-Ni solution, with the solid oxides tending to aggregate into clusters. As marked by red dot circles in Figure 6b, the liquid Cu-Ni solution started solidifying at 1161.8 °C, closely aligned with the liquidus temperature calculated through Thermo-Calc, as indicated in Figure 1. The number of oxide clusters significantly increased and spread on the remaining liquid Cu-Ni solution. When the temperature dropped to 1105.3 °C, as displayed in Figure 6c, the Cu-Ni solid phase significantly grew, and only a few spots of residual liquid Cu-Ni could be observed. The coating was fully solidified when the temperature was reduced to 1090.2 °C, as shown in Figure 6d.

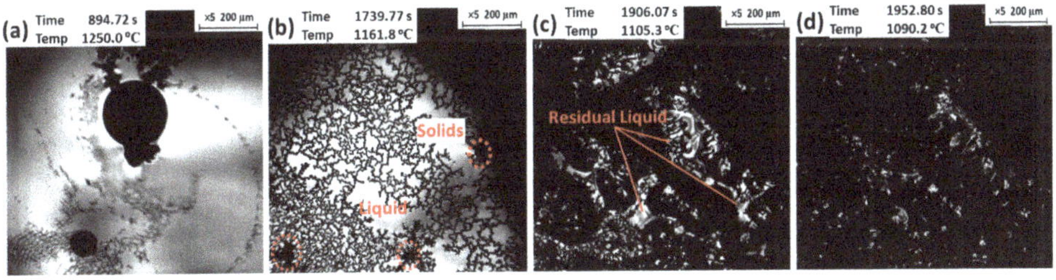

Figure 6. Solidification process of Cu-Ni coating. (**a**) Fully molten state, (**b**) solidification started, (**c**) solidified coating with less molten liquid, (**d**) fully solidified.

3.2. Surface Morphologies of the Coated Samples

Figure 7a,c,e present the overall surface morphologies of three different coated samples, and the red square regions in them were further investigated by secondary electron SEM imaging and displayed in Figure 7b,d,f. It can be found that the pure Cu powder-coated sample shows an undesirable convex surface shape, as seen in Figure 7a. In addition to the gold-shinny Cu appearance, a portion of the coating surface displays a dark color corresponding to the mixture of oxide and solidified Cu observed during HTCLSM. The micro-sized oxide particles can be found in Figure 7b.

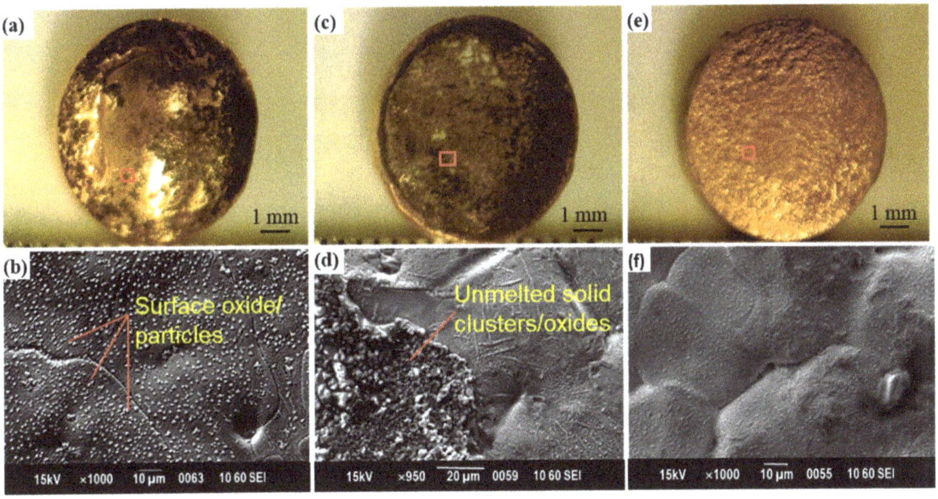

Figure 7. Surface morphologies and secondary electron images of coated samples, (**a**,**b**) pure Cu; (**c**,**d**) Cu–Cr; and (**e**,**f**) Cu-Ni.

As indicated in Figure 7c,d, the Cu–Cr coated sample's surface reveals a porous appearance and unmelted solid clusters/oxides, which is undesirable. In contrast, the surface of the Cu-Ni coated sample shown in Figure 7e,f appears smooth, and no dark region (oxidation). As shown in Figure 7f, the Cu-Ni coating layer displays fine and uniform polycrystalline structures, providing evidence that an isomorphous Cu-Ni solution was formed from the Cu and Ni powder composite after heating at 1250 °C. The Cu-Ni coated samples exhibit the most satisfactory surface appearance among the three coating combinations.

3.3. *Characterization of Coating Structure*

3.3.1. Coating Thickness and Defects

Figure 8a displays that the coating thickness deposited on the steel substrate by pure Cu powder is not uniform. The surface tension of the liquid Cu pool tends to drag the liquid Cu to the center of the round specimen, creating a convex layer of 960 μm thick in the center and significantly thinner, around 100 μm, at the edge. A substantial amount of casting porosities, up to 184.75 μm in radius, existed within the coated layer. The formation of porosity defects was attributed to the fact that the sample was heated much higher than the melting temperature of Cu, followed by rapid cooling. The existence of porosity defects significantly worsens the coating quality and can potentially reduce anti-corrosion performance when exposed to an aggressive environment [49]. As observed from the high-magnified view of the interface between the Cu coating layer and the steel substrate in Figure 8d, the Cu penetrated the grain boundary of the steel substrate to a depth of 35 μm.

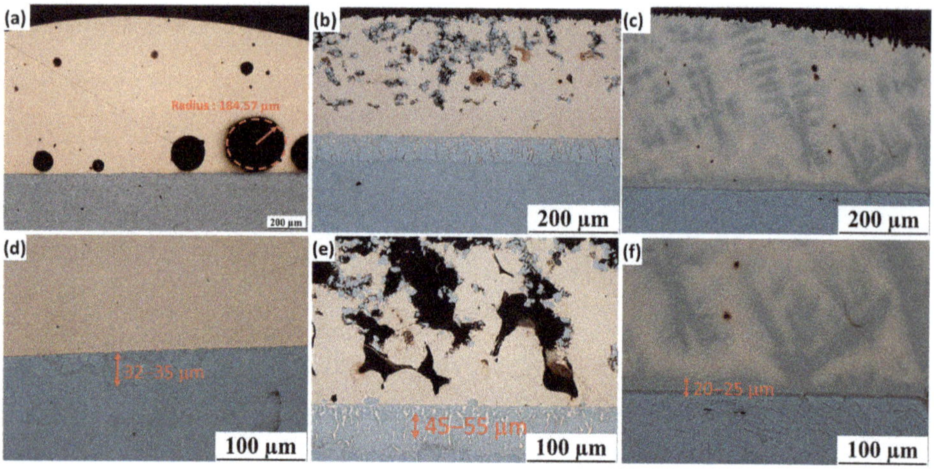

Figure 8. Cross-section of the coating layers at the center of coated samples, (**a**,**d**) pure Cu; (**b**,**e**) Cu–Cr; (**c**,**f**) Cu-Ni.

Despite the formation of a uniformly thick layer for the Cu–Cr coated sample, substantial cavities are visible throughout the coating, as indicated in Figure 8b and further confirmed in Figure 8e. These severe coating defects are caused by the high melting point of Cr and its insolubility in Cu [50]. Furthermore, Cu penetration into the grain boundaries of the steel substrate is up to 55 μm, which is even worse than the pure Cu-coated sample. This phenomenon will be confirmed and explained in the subsequent metallurgical examination.

In contrast, the Cu-Ni composite coatings exhibit a relatively uniform thickness distribution, ranging from 550 μm at the center to 350 μm at the edge, as shown in Figure 8c. It is interesting to report that a dendritic casting structure can be found in the Cu-Ni coating layer. It can be found in Figure 8f that a transition layer of 20–25 μm had formed between the Cu-Ni coating and the steel substrate. This transition layer is expected to be a Ni-rich region due to the high affinity of Ni towards Fe as compared to Cu's lack of miscibility with Fe. Ni plays an essential role in securing the bonding integrity of the coating and minimization of Cu penetration into the parent steel. A detailed metallurgical examination can be found in the following session.

3.3.2. Coating Structure and Integrity at Coating and Substrate Interface

While Cu is a primary element used to combat corrosion in the proposed metallic surface alloying technique due to its low cost, hot shortness can occur during the hot forming of Cu-containing steel. Therefore, introducing the Cr and Ni alloy elements and forming a diffused interface between the anti-corrosion metallic compound (instead of pure Cu) and the steel substrate is vital. This not only provides bonding strength but also suppresses Cu penetration into the parent steel.

Figure 9 shows distinct evidence of penetration and flow of liquid Cu into austenite grain boundaries of the Q235 substrate under pure Cu surface alloying. It is important to note that the mechanism of Cu enrichment for the surface alloying technique is different from that of Cu-containing steel. During the surface alloying, the Cu did not diffuse from the bulk steel and precipitated into the austenite grain boundaries of the steel. Instead, liquid Cu attracted the steel substrate and penetrated the austenite grain boundaries of the steel substrate. Figure 9b presents the EDS layered map and its corresponding Cu/Fe elemental maps. It can be found that the penetrated liquid Cu tended to detach steel grains from the metal surface. This phenomenon is known as liquid metal corrosion or liquid metal embrittlement [51,52]. Based on the reference [53], the penetration of liquid Cu into grain boundaries of substrate steel occurs rapidly. In general, although the Cu coating was successfully deposited on the steel surface, the significant Cu enrichment and penetration along grain boundaries of the steel substrate could trigger the hot shortness and surface cracking during the subsequent hot forming process. Hence, the surface alloying of pure Cu powder is undesirable due to the occurrence of liquid metal corrosion.

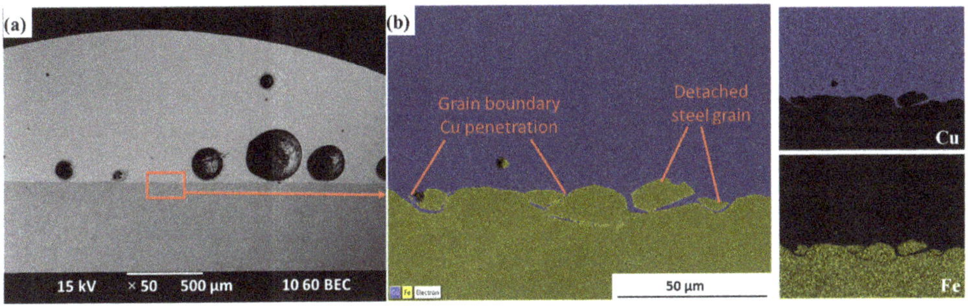

Figure 9. Pure Cu coating structure (**a**) Secondary Electron Image, (**b**) EDS layered map, and its Cu and Fe elemental maps.

As shown in Figure 10, the cross-section of the Cu–Cr coated sample consists of three distinct regions: the coating layer, diffusion zone, and substrate. It can be found that Cu and Cr elements did not form a homogeneous solution at the designed composition and thermal excursion history. The Cr particles remained unmelted and can still be observed within the Cu-based coating layer. The Fe element diffused from the steel matrix was combined with unmelted Cr particles to form a Cu-depleted Fe–Cr phase, as indicated in Figure 10a,b. These Fe–Cr phases tended to gather together, which was observed as the cluster in Cu–Cr coating. The presence of the Fe–Cr clusters introduced an inhomogeneous structure to the coating layer, potentially compromising the anti-corrosion performance.

More importantly, it is crucial to note that despite Cr diffusing into the steel matrix to form a Cr-rich transition zone around 45~55 μm between the coating layer and the substrate, it also facilitated Cu penetration into the grain boundaries, as indicated in Figure 10c,d. This phenomenon is attributed to the chemical actions among Fe, Cr, and Cu at the transition region. The Fe–Cu system is immiscible, and the same holds for the insolubility of Cu and Cr elements [50]. Consequently, the L(Cu), α(Fe,Cr), and γCu phases were expected to coexist during the surface alloy at the temperature range of 1100–1250 °C [54]. The insolubility of liquid Cu and α(Fe,Cr) resulted in the penetration of liquid Cu into the grain boundaries of the substrate. This observation confirms that Cu–Cr surface alloying is unable to prevent liquid Cu penetration, which could cause hot shortness in the subsequent forming process.

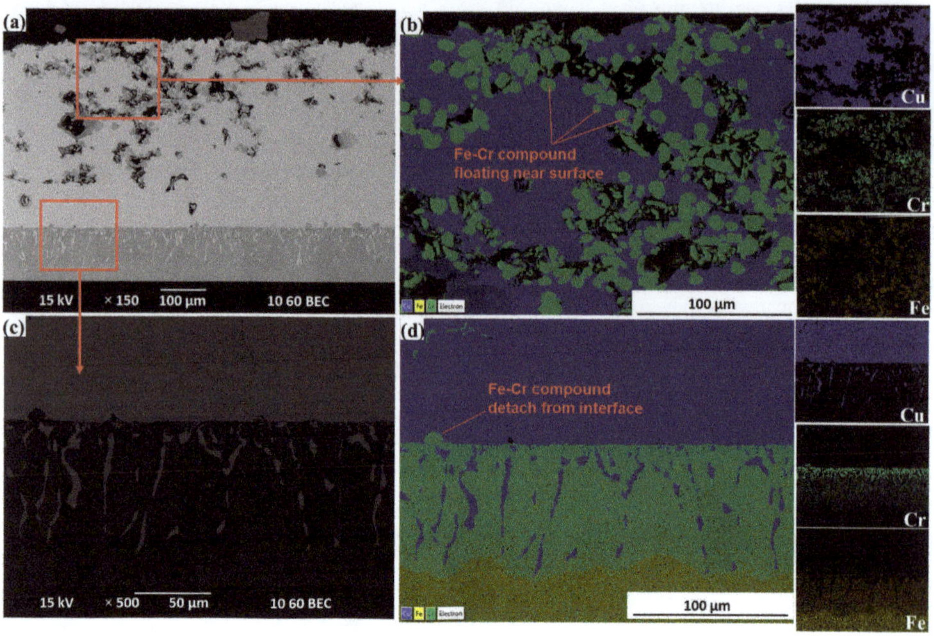

Figure 10. Cu–Cr coating structure (**a**) Back-scattered Electron Microscopic Image, (**b**) EDS layered map of the surface region. (**c**) Back-scattered electron image and (**d**) EDS layered map of the interface region. Cu, Cr, and Fe elemental maps for (**b**,**d**) are shown on the right side of respective EDS layered maps.

The microstructure of the metallic coating formed by the proposed Cu-Ni surface alloying techniques can be found in Figure 11. The sample exhibits three regions, including the cupronickel coating, a continuous binding transition layer between the cupronickel coating and the substrate, and the Q235 steel substrate. The Cu and Ni homogenously coexisted at the Cu-Ni coating, which confirmed the nature of the cupronickel solution alloy. As evidenced by elemental spectrums of the point scan in Figure 11a,e, the chemical composition of deposited cupronickel coating was 89.3 wt% Cu, 6.2 wt% Ni, and 4.5 wt% Fe. Compared with the initial metallic powder composite of 87.5 wt% Cu and 12.5 wt% Ni, the weight percentage of Ni is slightly reduced. As shown in Figure 11a, the dendritic phases were formed within the cupronickel coating. The elemental maps in Figure 11b display that the dendrites are rich in Ni and Fe and depleted in Cu, confirming the FeNi phase. At the elevated temperature of 1250 °C, the Fe could be dissolved into the cupronickel solution via two possible mechanisms, i.e., (1) Fe elements from the steel substrate could diffuse into the melted coating composite; and/or (2) the steel grains could be detached from the base metal by liquid metal corrosion [51]. Regardless of the source, the Fe element reacted with the Ni element during the solidification process due to the high affinity of Fe towards Ni as compared to Cu, resulting in the formation of a new phase that solidifies into a dendritic structure. The chemical composition of dendrites consisted of 62.7% Cu, 19.3 wt% Ni, and 18.1 wt% Fe, as seen in Figure 11f. It is important to note that Cu concentrations remained high in both of the cupronickel solutions or dendrites. At the same time, the Ni percentage is comparatively higher in dendrites when compared to the rest of the coating.

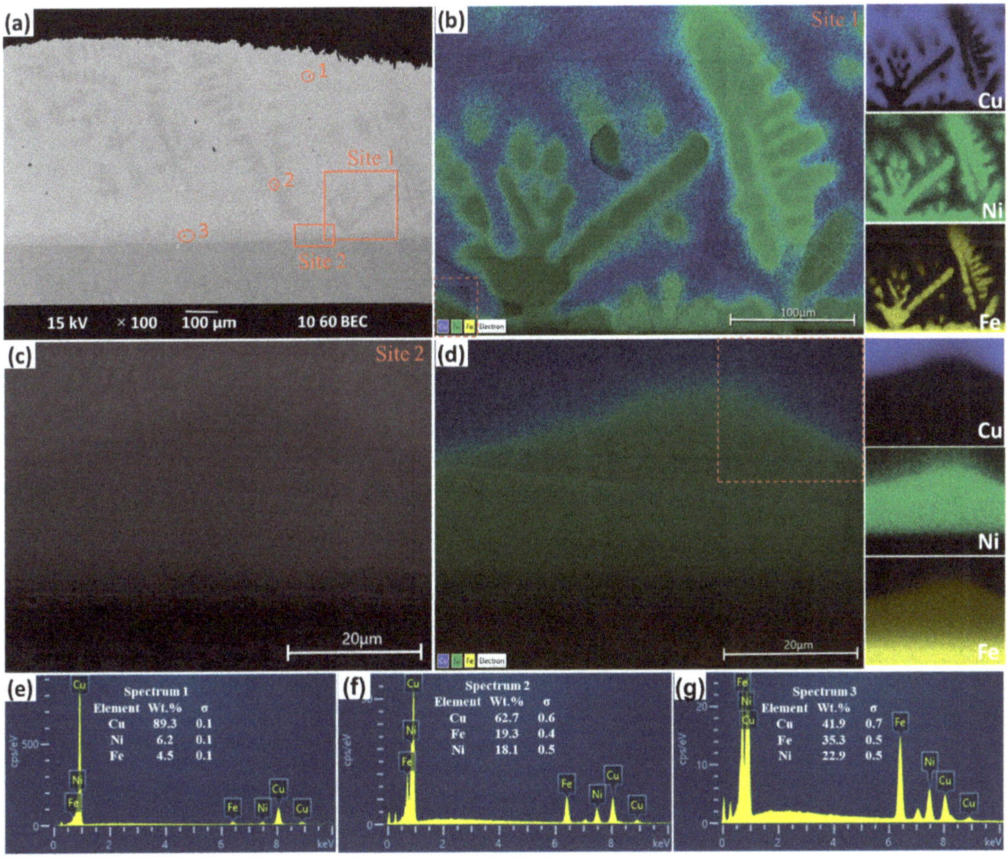

Figure 11. Cu-Ni coating structure (**a**) Back-scattered Electron Image, (**b**) EDS layered map and Elemental maps of Cu, Ni, and Fe at Site 1 (Ni-rich dendritic structure), (**c**) Back-scattered Electron Image of Site 2 (Ni-rich interface); (**d**) EDS layered map and Elemental maps of Cu, Ni, and Fe at Site 2 (Ni-rich interface); (**e–g**) Elemental spectrums of Point scan 1 (coating region), (**f**) point scan 2 (dendrite), (**g**) point scan 3 (interface). In (**a**), an overlapping region between Site 1 and Site 2 is highlighted by the red dashed rectangular boxes in the EDS mapping shown in (**b**,**d**).

Although the thickness of the cupronickel coating is up to 500 μm, as shown in Figure 11a, the thickness of the binding transition layer between the cupronickel coating and the substrate is around 20–25 μm. The EDS mapping results depicted in Figure 11d confirm that the binding transition layer is enriched with Ni. The chemical composition of the Ni-rich transition layer was analyzed through elemental spectrum scanning in Figure 11g, revealing a composition of 41.9 wt% Cu, 22.9 wt% Ni, and 35.3 wt% Fe. The formation of the Ni-rich transition layer and the high percentage of Cu in the Ni-rich transition layer align with the expected behavior of a ternary Cu-Ni–Fe system. Both liquid L Cu-Ni–Fe and solid γ Cu-Ni–Fe phases coexist when a Cu-Ni–Fe system is exposed to high temperatures. The L phase with a lower Ni/Cu ratio flows upwards to form the cupronickel coat. The γ Cu-Ni–Fe solution phases with a higher Ni/Cu ratio are more enriched on an interface to promote the Ni-rich transition layer. It is interesting to report that even though the Cu concentrations are different in the coating and transition layer, there are still relatively high Cu concentrations of 41.9 wt% in the Ni-rich transition layer [13]. The Ni-rich interface layer is continuous and adhesive, effectively preventing

Cu penetration into the substrate grain boundaries, which is a desirable characteristic in surface alloying.

It is intriguing to compare the characteristics of the transition layers formed by the Cu–Cr and Cu-Ni composites. Rothman et al. [55] measured the simultaneous lattice diffusion of Fe, Cr, and Ni, revealing that the diffusion coefficient of Cr is 1.39 times that of Fe and 2.5 times of Ni. Consequently, the Cr can diffuse deeper into the steel matrix compared to Ni. It aligns well with current research that the Cr-rich transition zone formed by the Cu–Cr coating has a thickness ranging from 45–55 μm, almost double the thickness of the Ni-rich transition layer observed in the Cu-Ni coating (20–25 μm). However, despite its thickness, the Cr-rich transition layer is unable to prevent Cu penetration during surface alloying at high temperatures. In contrast, the Ni-rich transition layer is capable of minimizing the risk of Cu penetration.

The material characterization confirms that the surface alloying technique utilizing Cu-Ni powder at high temperature successfully forms a coating comprising cupronickel solution and FeNi dendrites. In addition, a Ni-rich transition layer acts as a chemical binding between the cupronickel coating and the steel substrate, which plays a critical role in reducing liquid Cu penetration at the steel grain boundary to minimize the risk of hot shortness during subsequent metal forming.

The significant difference in coating performance results from the types of bond interaction between Cr and Ni in the structures of Cu–Cr and Cu-Ni alloys. Copper and chromium have different atomic radii and crystalline structures (Cu has an fcc structure, while Cr has a bcc structure). In contrast, nickel is similar to copper in terms of atomic size and crystal structure, with both having fcc structures. The electronegativity difference between Cu and Cr is greater than that between Cu and Ni, leading to stronger metallic bonds in Cu-Ni alloys and some covalent character in Cu–Cr alloys. The similar atomic sizes and electronegativity and the identical fcc structures of Cu and Ni facilitate the formation of solid solutions. In contrast, the differing atomic sizes, electronegativity, and crystalline structures of Cu and Cr result in limited solubility and the formation of distinct phases.

In future studies, it will be beneficial to investigate the optimized structures of Fe–Cr and compare the stabilized energies of Fe–Cr, Cu–Cr, and Cu-Ni at the temperatures of interest.

3.4. Coating Formation in Practical Hot Rolling and Corrosion Resistance Performance

To evaluate the proposed surface alloy technique, a series of hot rolling experiments were conducted using a TMAX tube furnace manufactured by Xiamen Tmax Battery Equipments Limited, Xiamen, China, and an experimental Hille 100 rolling mill manufactured by Hille Engineering Company Ltd., Sheffield, UK.

The data collected from the high-temperature confocal laser scanning microscopy indicated a suitable temperature range of 1099.5–1175 °C for melting the mixture of 87.5 wt% Cu and 12.5 wt% Ni powders. Hence, a heating temperature of 1150 °C was selected to mimic the actual reheating condition of an industry hot strip rolling mill. The process involved covering 20 mm thick Q235 steel slabs with Cu-Ni metallic powders, which were then heated in a tube furnace filled with argon gas to 1150 °C and held at that temperature for 25 min to facilitate the surface alloying technique. Subsequently, a 20% reduction was applied to the heated slabs immediately after removing them from the furnace. The rolled samples were air-cooled to room temperature.

Figure 12a displays a representative cross-section image of the coated slab after hot rolling, revealing the formation of a compact metallic coating with an average thickness of 220 μm. The line-scan analysis in Figure 12b shows a sharp increase in copper intensity from the transition region, which remains high throughout the coating layer. The intensities of Ni and Fe were nearly identical across the coating layer, aligning with HTCLSM's results in Figure 11e, consisting of nearly the same amount of 6.2 wt% Ni and 4.5 wt% Fe in the coating layer. Fe intensity increases gradually towards the steel matrix, while Ni tends to concentrate at the interface.

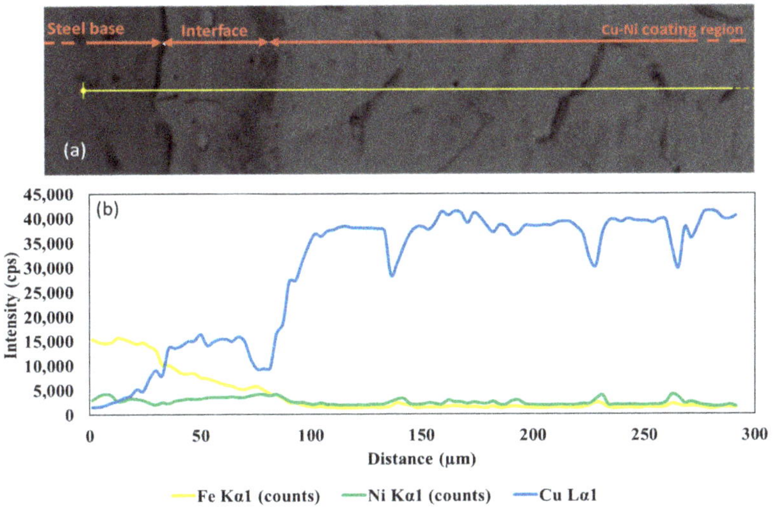

Figure 12. (**a**) Cu-Ni coated sample cross-section after practical hot rolling and (**b**) line-scan of the cross-section.

To investigate the effectiveness of cupronickel-coated steel in terms of anti-corrosion performance, a series of electrochemical corrosion tests were conducted in a 3.5% NaCl solution using a Gamry electrochemical corrosion testing rig over a period of 10 min, 1 day, 7 days, 14 days, and 30 days. The tests were performed on both cupronickel-coated steel and uncoated bare steel for comparison. Figure 13 presents the typical polarization curves after 1 week of salted water immersion. The results confirmed the superior corrosion-resistive properties of the Cu-Ni coating compared to uncoated Q235 steel. The corrosion rate of the coated sample was calculated to be 33.42 μm/yr., which is 44% lower than that of the uncoated sample, recorded to be 59.78 μm/yr. These findings indicate a significant improvement in corrosion resistance with the application of the Cu-Ni coating.

In summary, in situ thermal interactions of three composites, including pure Cu powder, a mixed powder of 87.5 wt% Cu-12.5 wt% Ni, and a mixed powder of 90 wt% Cu-10 wt% Cr, were tested using a HTCLSM, considering the maximum hot working temperatures of steel in the industry (1200–1250 °C). The Cu-Ni powder mixture began to melt at 1099.5 °C and was fully melted at 1175 °C, significantly different from the Cu-Ni solid solution and bulk Cu or Ni. As a result of high-temperature reactions, copper penetration of up to 35 μm for pure copper and 55 μm for copper-chromium composite coatings occurred due to liquid metal corrosion. In contrast, the Cu-Ni metallic powder combination ensures a cupronickel solution coating, with a chemical composition of 89.3 wt% Cu, 6.2 wt% Ni, and 4.5 wt% Fe, which inhibits copper penetration. The cupronickel coating presents uniform thickness, superior surface morphology, and continuous coverage. A 44% improvement in corrosion resistance was confirmed by electrochemical corrosion tests.

To achieve an effective corrosion-resistant layer, current research suggests optimizing the specific components of Cu and Ni, with a composite ratio of 7:1 (Cu:Ni). However, future studies should explore the synergistic effects of multi-element alloying, including Cr, Ni, Cu, and others, to further enhance corrosion resistance. The detailed electrochemical corrosion tests on various Cu, Cu–Cr, and Cu-Ni composite coatings, along with the impact of uneven coatings and the presence of defects on corrosion resistance, will be addressed in a forthcoming report.

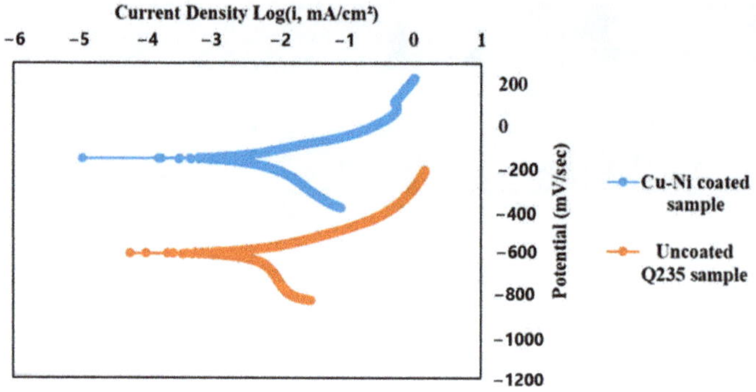

Figure 13. Electrochemical corrosion test results in 3.5% salt solution.

4. Conclusions

A high-temperature surface alloying technology was developed to implement the Cu-based metallic coating on carbon steel surfaces to improve corrosion resistance.

The metallic coatings could be formed on the steel surface due to a high-temperature reaction. However, the pure Cu and Cu–Cr metallic powder combinations were unsuitable due to excessive Cu penetration. The Cu-Ni metallic powder combination demonstrated a high coating quality providing full surface coverage, a consistent coating thickness, and no Cu penetration into grain boundaries.

The surface alloying using a Cu-Ni composite comprised a cupronickel solution coat with FeNi dendrites and a Ni-rich transition layer. The presence of the Ni-rich transition layer is of significant importance as it inhibited Cu from penetrating the steel matrix grain boundary while simultaneously enhancing the adhesion between the coating and the substrate.

The integrity of the coating was further confirmed by hot rolling experiments. The electrochemical corrosion test proved that the Cu-Ni coating exhibited a significantly higher corrosion resistance, with a 44% improvement compared to uncoated Q235 steel.

Author Contributions: Conceptualization, H.Z. and W.W.; methodology, H.H.K., T.W., H.L., L.S.; software, T.W.; validation, H.H.K. and Q.Z.; formal analysis, H.H.K. and H.Z.; investigation, H.H.K., L.S. and H.Z.; resources, A.Y., Z.L.; writing—original draft preparation, H.H.K. and H.Z.; writing—review and editing, H.H.K. and H.Z.; supervision, H.L. and H.Z.; project administration, A.Y., Z.L., W.W. and H.Z.; funding acquisition, A.Y., Z.L., W.W. and H.Z. All authors have read and agreed to the published version of the manuscript.

Funding: This research was funded by Baosteel-Australia Joint Research and Development Centre (BA17012).

Institutional Review Board Statement: Not applicable.

Informed Consent Statement: Not applicable.

Data Availability Statement: Data will be made available on request.

Acknowledgments: The authors acknowledge the use of FEI Nanolab G3 CX at the UOW Electron Microscopy Centre, which is funded by the Australian Research Council (ARC)—Linkage, Infrastructure, Equipment and Facilities (LIEF) grant (LE160100063). The author, Khan, acknowledges the support of HEC-Pakistan.

Conflicts of Interest: The authors Huda Hanif Khan, Tong Wang, Lihong Su, Huijun Li, Qiang Zhu, Hongtao Zhu declare no conflict of interest. The authors Ana Yang, Zigang Li, Wei Wang are currently employed by Baowu Steel Group Corporation and were employed by Baowu Steel Group Corporation while contributing to this manuscript. Their contributions to this work and manuscript were approved by Baowu Steel Group Corporation, and this manuscript was reviewed and approved by Baowu Steel Group Corporation. They received no financial compensation from any source for the contributions they made to the scientific work and manuscript.

erences

Wang, Y.; Liu, W.; Li, N.; Zheng, C. The Importance of Structure and Corrosion Resistance of Steels/Alloys. *Coatings* **2022**, *12*, 997. [CrossRef]

Exarchos, D.A.; Sima, A.; Ananiadis, E.A.; Dalla, P.T.; Tragazikis, I.K.; Kordatou, T.Z.; Farmaki, S.; Gontzamanis, A.; Kechagias, V.; Kechagias, L.; et al. Development and characterization of environmentally friendly multifunctional protective coatings. *Proc. SPIE* **2022**, *12049*, 102–116.

Shevchenko, V.Y.; Shilova, O.A.; Kochina, T.A.; Barinova, L.D.; Belyi, O.V. Improving the Safety of the Transportation System and Resource Conservation through the Introduction of Environmentally Safe Protective Coatings. *Glass Phys. Chem.* **2019**, *45*, 1–9. [CrossRef]

Bangaru, N.V.; Krutenat, R.C. Diffusion coatings of steels: Formation mechanism and microstructure of aluminized heat-resistant stainless steels. *J. Vac. Sci. Technol. B Microelectron. Process. Phenom.* **1984**, *2*, 806–815. [CrossRef]

Wang, Z.B.; Lu, J.; Lu, K. Chromizing behaviors of a low carbon steel processed by means of surface mechanical attrition treatment. *Acta Mater.* **2005**, *53*, 2081–2089. [CrossRef]

Wang, D.; Shi, Z. Aluminizing and oxidation treatment of 1Cr18Ni9 stainless steel. *Appl. Surf. Sci.* **2004**, *227*, 255–260. [CrossRef]

Kula, P.; Pietrasik, R.; Dybowski, K. Vacuum carburizing—Process optimization. *J. Mater. Process. Technol.* **2005**, *164–165*, 876–881. [CrossRef]

Sun, M.; Du, C.; Liu, Z.; Liu, C.; Li, X.; Wu, Y. Fundamental understanding on the effect of Cr on corrosion resistance of weathering steel in simulated tropical marine atmosphere. *Corros. Sci.* **2021**, *186*, 109427. [CrossRef]

Zhang, J.; Zhang, W.; Wei, L.; Pu, L.; Liu, J.; Liu, H.; Li, Y.; Fan, J.; Ding, T.; Guo, Z. Alternating Multilayer Structural Epoxy Composite Coating for Corrosion Protection of Steel. *Macromol. Mater. Eng.* **2019**, *304*, 1900374. [CrossRef]

Anagri, A.; Baitukha, A.; Debiemme-Chouvy, C.; Lucas, I.T.; Pulpytel, J.; Tran, T.T.M.; Tabibian, S.; Arefi-Khonsari, F. Nanocomposite coatings based on graphene and siloxane polymers deposited by atmospheric pressure plasma. Application to corrosion protection of steel. *Surf. Coat. Technol.* **2019**, *377*, 124928. [CrossRef]

Choi, Y.-S.; Shim, J.-J.; Kim, J.-G. Effects of Cr, Cu, Ni and Ca on the corrosion behavior of low carbon steel in synthetic tap water. *J. Alloys Compd.* **2005**, *391*, 162–169. [CrossRef]

Salter, W.J.M. Discussion on surface hot shortness of mild steel. *J. Iron Steel Inst.* **1962**, *200*, 750–751.

Akamatsu, S.; Senuma, T.; Takada, Y.; Hasebe, M. Effect of nickel and tin additions on formation of liquid phase in copper bearing steels during high temperature oxidation. *Mater. Sci. Technol.* **1999**, *15*, 1301–1307. [CrossRef]

Yamamoto, K.I.; Toh, T.; Hamatani, H.; Tsunenari, K.; Umetsu, K.; Maruki, Y.; Takeuchi, S.; Yamada, Y. Development of steel surface melting technology for improvement of hot shortness caused by tramp elements. *Nippon Steel Tech. Rep.* **2013**, *104*, 69–73.

Uhlig, H.H. Passivity in Copper-Nickel and Molybdenum-Nickel-Iron Alloys. *Trans. Electrochem. Soc.* **1944**, *85*, 307. [CrossRef]

Ezuber, H.; Al-Shater, A.; Murra, F.; Al-Shamri, N. Corrosion Behavior of Copper-Nickel Alloys in Seawater Environment. In Proceedings of the 16th Middle East Corrosion Conference & Exhibition, Manama, Bahrein, 8–11 February 2016.

Tolulope Loto, R. Correlative investigation of the corrosion susceptibility of C70600 and C26000 copper based alloys for application in seawater environment. *Mater. Today Proc.* **2022**, *65*, 2151–2155. [CrossRef]

Qian, Y.H.; Niu, D.; Xu, J.J.; Li, M.S. The influence of chromium content on the electrochemical behavior of weathering steels. *Corros. Sci.* **2013**, *71*, 72–77. [CrossRef]

Park, S.A.; DLe, P.; Kim, J.G. Alloying Effect of Chromium on the Corrosion Behavior of Low-Alloy Steels. *Mater. Trans.* **2013**, *54*, 1770–1778. [CrossRef]

Hosmani, S.S.; Kuppusami, P.; Goyal, R.K. (Eds.) Basics of Surface Alloying. In *An Introduction to Surface Alloying of Metals*; Springer: New Delhi, India, 2014; pp. 1–28.

Ezaki, H.; Nambu, T.; Ninomiya, R.; Nakahara, Y. Estimation of liquidus temperature of Sn-based alloys and its application to the design of Pb-free solder. *J. Mater. Sci. Mater. Electron.* **2002**, *13*, 269–272. [CrossRef]

Djurdjevic, M.; Manasijević, S.; Odanović, Z.; Dolić, N. Calculation of Liquidus Temperature for Aluminum and Magnesium Alloys Applying Method of Equivalency. *Adv. Mater. Sci. Eng.* **2013**, *2013*, 170527. [CrossRef]

Jia, Y.; Cheng, H.; Qiu, J.; Han, F. Effect of temperature on diffusion behavior of Te into nickel. *J. Nucl. Mater.* **2013**, *441*, 372–379. [CrossRef]

Ling, Z.; Wang, M.; Kong, L. Liquid Metal Embrittlement of Galvanized Steels During Industrial Processing: A Review. In *Transactions on Intelligent Welding Manufacturing*; Springer: Singapore, 2018; pp. 25–42.

Jeon, W.-S.; Sharma, A.; Jung, J.P. Liquid Metal Embrittlement of Galvanized TRIP Steels in Resistance Spot Welding. *Metals* **2020**, *10*, 787. [CrossRef]

Ikeda, Y.; Yuan, R.; Chakraborty, A.; Ghassemi-Armakiet, H. Early stages of liquid-metal embrittlement in an advanced high-strength steel. *Mater. Today Adv.* **2022**, *13*, 100196. [CrossRef]

Liu, T.; Long, M.; Chen, D.; Wu, S.; Tang, P.; Liu, S.; Duan, H.; Yang, J. Investigations of the peritectic reaction and transformation in a hypo-peritectic steel: Using high-temperature confocal scanning laser microscopy and differential scanning calorimetry. *Mater. Charact.* **2019**, *156*, 109870. [CrossRef]

Ikhmayies, S. Using thermo-calc software to produce the phase diagram of Zn-Te system. *J. Energy Syst.* **2020**, *4*, 88–95. [CrossRef]

Shah, S.; Ullah, N.; Ullah, B.; Khan, M.S. Thermodynamic Analysis and Calculations of (Fe-Co) Alloy by Modeling and Simulation using Thermo-Calc Software. *J. Heterocycl.* **2018**, *1*, 35–38. [CrossRef]

30. Chikama, H.; Shibata, H.; Emi, T.; Susuki, M. "In-situ" real time observation of planar to cellular and cellular to dendri transition of crystals growing in Fe–C alloy melts. *Mater. Trans. JIM* **1996**, *37*, 620–626. [CrossRef]
31. Ren, Y.; Zhu, P.; Ren, C.; Liu, M. Dissolution of SiO2 Inclusions in CaO-SiO2-Based Slags In Situ Observed Using High-Temperatu Confocal Scanning Laser Microscopy. *Metall. Mater. Trans. B* **2022**, *53*, 682–692. [CrossRef]
32. Michelic, S.; Goriupp, J.; Feichtinger, S.; Kang, Y.B.; Bernhard, C.; Schenk, J. Study on Oxide Inclusion Dissolution in Seconda Steelmaking Slags using High Temperature Confocal Scanning Laser Microscopy. *Steel Res. Int.* **2016**, *87*, 57–67. [CrossRef]
33. Böhme, N.; Hauke, K.; Dohrn, M.; Neuroth, M.; Geisler, T. High-temperature phase transformations of hydroxylapatite and t formation of silicocarnotite in the hydroxylapatite–quartz–lime system studied in situ and in operando by Raman spectroscopy *Mater. Sci.* **2022**, *57*, 15239–15266. [CrossRef]
34. Branch, S.D.; Felmy, H.M.; Medina, A.S.; Bryan, S.A.; Lines, A.M. Exploring the Complex Chemistry of Uranium within Molt Chloride Salts. *Ind. Eng. Chem. Res.* **2023**, *62*, 14901–14909. [CrossRef]
35. Zhang, B.; Tekle, H.; O'Malley, R.J.; Sander, T.; Smith, J.D.; Gerald II, R.E.; Huang, J. In Situ and Real-Time Mold Flux Analy Using a High-Temperature Fiber-Optic Raman Sensor for Steel Manufacturing Applications. *J. Light. Technol.* **2023**, *41*, 4419–44 [CrossRef]
36. Zhang, B.; Tekle, H.; O'Malley, R.J.; Smith, J.D.; Gerald II, R.E.; Huang, J. In Situ High-Temperature Raman Spectroscopy vi Remote Fiber-Optic Raman Probe. *IEEE Trans. Instrum. Meas.* **2023**, *72*, 6002008. [CrossRef]
37. Galedari, S.A.; McDonald, A. A Comprehensive Review of Corrosion Resistance of Thermally-Sprayed and Thermally-Diffus Protective Coatings on Steel Structures. *J. Therm. Spray Technol.* **2019**, *28*, 645–677. [CrossRef]
38. Oskay, C.; Meisner, T.M.; Dobler, C.; Grégoire, B.; Galetz, M.C. Scale Formation and Degradation of Diffusion Coatings Deposit on 9% Cr Steel in Molten Solar Salt. *Coatings* **2019**, *9*, 687. [CrossRef]
39. Boulesteix, C.; Pedraza, F. Characterisation of aluminium diffusion coatings elaborated on austenitic stainless steels and ferritic-martensitic steels. *Surf. Coat. Technol.* **2018**, *339*, 27–36. [CrossRef]
40. Grinon-Echaniz, R.; Refait, P.; Jeannin, M.; Sabot, R.; Paul, S.; Thornton, R. Study of cathodic reactions in defects of thermal spr aluminium coatings on steel in artificial seawater. *Corros. Sci.* **2021**, *187*, 109514. [CrossRef]
41. Wu, J.; Zhang, S.D.; Sun, W.H.; Wang, J.Q. Influence of oxidation related structural defects on localized corrosion in HVAF-spray Fe-based metallic coatings. *Surf. Coat. Technol.* **2018**, *335*, 205–218. [CrossRef]
42. Takeuchi, S.; Ito, M.; Takeda, K. Modelling of residual stress in plasma-sprayed coatings: Effect of substrate temperature. *Su Coat. Technol.* **1990**, *43–44*, 426–435. [CrossRef]
43. *ASTM A588/A588M-19*; Standard Specification for High-Strength Low-Alloy Structural Steel, up to 50 ksi [345 MPa] Minimu Yield Point, with Atmospheric Corrosion Resistance. ASTM: West Conshohocken, PA, USA, 2019; p. 3.
44. Vattur Sundaram, M.; Karamchedu, S.; Gouhier, C.; Hryha, E. Vacuum sintering of chromium alloyed powder metallurgy ste *Met. Powder Rep.* **2019**, *74*, 244–250. [CrossRef]
45. Morcillo, M.; Chico, B.; Cano, H.; de la Fuente, D. Weathering steels: From empirical development to scientific design. A revi *Corros. Sci.* **2014**, *83*, 6–31. [CrossRef]
46. Wang, W. Review on studies of reducing Cu-induced surface hot shortness in steel. *Baosteel Tech. Res.* **2015**, *9*, 3–13.
47. Servant, C.; Sundman, B.; Lyon, O. Thermodynamic assessment of the Cu Fe Ni system. *Calphad* **2001**, *25*, 79–95. [CrossRef]
48. Zhao, Y.; Yu, Z.; Planche, M.P.; Lasalle, A. Influence of Substrate Properties on the Formation of Suspension Plasma Spray Coatings. *J. Therm. Spray Technol.* **2018**, *27*, 73–83. [CrossRef]
49. Zhang, S.D.; Wu, J.; Qi, W.B.; Wang, J.Q. Effect of porosity defects on the long-term corrosion behaviour of Fe-based amorpho alloy coated mild steel. *Corros. Sci.* **2016**, *110*, 57–70. [CrossRef]
50. Ma, E. Alloys created between immiscible elements. *Prog. Mater. Sci.* **2005**, *50*, 413–509. [CrossRef]
51. Zhang, J.; Hosemann, P.; Maloy, S. Models of liquid metal corrosion. *J. Nucl. Mater.* **2010**, *404*, 82–96. [CrossRef]
52. Balbaud-Célérier, F.; Courouau, J.L.; Feron, D.; Martinell, L. Corrosion of structural materials by liquid metals used in fusi fission, and spallation. In *Nuclear Corrosion*; Ritter, S., Ed.; Woodhead Publishing: Sawston, UK, 2020; pp. 437–457.
53. Shibata, K.; Seo, S.J.; Kaga, M.; Uchino, H.; Sasanuma, A.; Asakura, K.; Nagasaki, C. Suppression of surface hot shortness due Cu in recycled steels. *Mater. Trans.* **2002**, *43*, 292–300. [CrossRef]
54. Dreval', L.; Turchanin, M.A.; Abdulov, A.R.; Bondar, A.A. Thermodynamic assessment of the Cu–Fe–Cr phase diagram. *Che Met. Alloys* **2010**, *3*, 132–139. [CrossRef]
55. Rothman, S.; Nowicki, L.; Murch, G. Self-diffusion in austenitic Fe-Cr-Ni alloys. *J. Phys. F Met. Phys.* **1980**, *10*, 383. [CrossRef

coatings

MDPI

icle

esearch on Drag Reduction by Coating the Inner Wall of ydraulic Pipeline

Wang [1], Junjie Zhou [1,2,*], Wenbo Liao [1] and Shihua Yuan [1]

1 School of Mechanics and Vehicles, Beijing Institute of Technology, Beijing 100081, China; 3120185221@bit.edu.cn (X.W.); bit10007liao@163.com (W.L.); bit10007zhou@163.com (S.Y.)
2 Institute of Frontier Technology, Beijing Institute of Technology, Beijing 250101, China
* Correspondence: bit_zhou50082@163.com

Abstract: This study employs computational fluid dynamics (CFD) simulations to investigate the effect of wall roughness on linear loss in circular pipelines. It specifically addresses hemispherical roughness, focusing on how changes in spacing influence linear loss, a critical determinant of fluid motion within pipelines. The simulations further assess the impact of these variables on flow characteristics, laying a theoretical groundwork for drag reduction and pipeline design improvement. Results indicate that increased spacing between roughness elements reduces the differential pressure at both pipeline ends. The dimensionless spacing value of 30 stabilizes this pressure, suggesting a limit to further changes. Additionally, a rise in roughness height at this spacing exacerbates differential pressure, highlighting a proportional relationship between roughness dimensions and linear loss—greater roughness leads to higher linear loss. Applying a nickel-plated coating on the inner wall significantly lowers roughness, thereby reducing linear loss.

Keywords: roughness; CFD numerical simulation; linear loss; drag reduction

1. Introduction

The influence of wall roughness on boundary layer flow within pipelines has garnered significant academic interest, contributing extensively to the body of research on single-phase turbulent flow through channels with various roughness geometries [1–3]. Research has extensively explored how different roughness shapes affect fluid flow in pipelines [4]. Oktarina et al. [5] investigated the impact of cylindrical surface roughness on fluid flow characteristics, finding that increased roughness disrupts flow and enhances vortex formation at greater distances. Similarly, Ashmawy [6] studied the effects of surface roughness on stress in fluid flows, demonstrating a marked decrease in flow velocity as longitudinal corrugation roughness increases. Yu et al. [7] analyzed flow resistance in pipelines, highlighting vortices induced by surface roughness and introduced a new method for calculating the friction coefficient. Memento et al. [8] performed experiments to determine how pipeline diameter and surface roughness influence the friction coefficient in fluid flows, noting that both inlet conditions and surface roughness impact the friction coefficient across different pipeline diameters. They also observed a significant increase in pressure drop magnitude with decreasing pipeline size. Further, Oktarina et al. [9] examined the effects of wall roughness on resistance to fluid transport in nanopores, revealing a more pronounced impact in smaller nanotubes. Du et al. [10] used scanning electron microscopy (SEM) to create a three-dimensional model of surface roughness and explored its effect on flow resistance in microtubes, finding that increased roughness significantly raises flow resistance, especially when relative roughness exceeds 3%. Yang et al. [11] investigated flow resistance along rough boundaries in open channels and pipelines, affirming the boundary shear stress summation principle's applicability, which consistently aligns with experimental data in depicting the friction coefficient in pipelines. Lastly, Song et al. [12]

studied the impact of surface roughness on flow fields and pressure drops in circular pipelines, developing a theoretical model to quantify roughness effects on fluid dynamics.

Research on the impact of small hemispherical roughness, whether regularly or randomly distributed, on channel walls remains sparse. Wu et al. [13] performed direct numerical simulations (DNS) of turbulent flows over hexagonally packed hemispheres, establishing a scaling relationship between universal wall shear stress variance and average size, which supports the development of wall models for large eddy simulations (LES). Iyer et al. [14] investigated the effects of discrete and distributed roughness through DNS, finding that roughness can induce coherent streamwise vortices and enhance fluid transition. Philip [15] conducted experimental studies on single hemispherical roughness on a flat plate within a zero-pressure gradient laminar boundary layer, focusing on the Strouhal number-related behavior of "hairpin" vortices induced by roughness, their role in the development of a turbulent boundary layer, and their dependency on inflection point instability. Zhou et al. [16] combined DNS with the immersed boundary method to explore boundary layer flow transition induced by three-dimensional roughness, revealing the evolution of hairpin vortices and secondary vortex structures. A subsequent simulation demonstrated quantitative consistency with experimental data using two-component Particle Image Velocimetry (PIV). Michele et al. [17] examined vortex structure in turbulent boundary layers over sparse hemispherical roughness, confirming the presence of vortex packets in the outer flow layer. Prahladh [18] utilized DNS to study the transition from laminar to turbulent flow in high-speed laminar boundary layers caused by discrete hemispherical roughness. Observations indicated that flow with a Mach number of 8.23 remained laminar downstream of the roughness, while transitions occurred at lower Mach numbers. Caviezel et al. [19] conducted DNS and LES of fully developed turbulent flow in channels with smooth-walled and hemispherical roughness at shear Reynolds numbers (Re) of 180 and 400, finding that the friction factor decreases with increasing Reynolds number and roughness spacing but increases significantly with roughness height. Interestingly, the random distribution of these cells had a relatively minor effect on both the friction factor and average velocity. Additionally, strategic placement of roughness facilitated immediate lateral movement of flow within the wall layer, effectively enhancing fluid transport. Qin et al. [20] performed molecular dynamics simulations to assess the effects of various roughness shapes, including hemispherical, on nanofluid flow in nanochannels. Results indicated that surface roughness reduces the range of density fluctuation near nanochannel walls, with triangular roughness having the most significant impact on nanofluid flow characteristics.

Regarding fluid transport in pipelines, linear loss is a critical factor that cannot be overlooked, primarily stemming from the interaction between the fluid and the pipeline wall [21]. This interaction significantly impacts the efficiency and performance of the fluid conveyance system. Recognizing the importance of minimizing linear losses is essential for the design and operation of fluid transport systems [22]. Optimization of pipeline design, including adjustments to pipeline diameter and surface roughness, along with the implementation of drag reduction devices [23], can facilitate this. Such measures not only enhance system energy efficiency and reduce operational costs but also extend the lifespan of pipelines and decrease maintenance requirements [24]. Consequently, this study employs numerical simulation to examine the effects of the height and spacing of hemispherical roughness, regularly distributed along channel walls, on linear loss. The objective is to diminish drag by reducing wall roughness [25].

This paper investigates the impact of inner wall roughness on the drag reduction performance of turbulent hydraulic oil flow (Re > 5000) in hydraulic steel pipelines using computational fluid dynamics (CFD) for numerical simulation. The study focuses on the enhancement of drag reduction by applying an inner wall surface coating, thereby reducing roughness. Initially, hemispherical roughness units are modeled on the inner surface of the circular pipe using a 1/4 circular pipeline model to decrease computational demands. The study also explores how variations in roughness spacing influence drag reduction, thereby advancing our understanding of roughness in pipeline hydrodynamics. Furthermore, a nickel-plated coating is applied to the inner wall to reduce roughness and minimize linear losses, providing insights for optimizing pipeline design and reducing fluid resistance. This investigation builds on findings from Reference [26], which demonstrated that a nickel-plated metal coating on the inner walls of hydraulic steel pipelines effectively reduces linear loss in laminar flow. Extending these findings, this study explores turbulence reduction by applying a similar nickel-plated coating, aiming to decrease wall roughness in hydraulic steel pipelines. The numerical simulations assess the influence of coating thickness variations on roughness modification and the consequent drag reduction, explaining these effects through changes in the height and morphology of the roughness.

2. Numerical Simulation of Pipeline Drag Reduction

This study employs three-dimensional numerical simulations to explore turbulent fluid flow through horizontally oriented, straight circular pipelines using Ansys Fluent software. The Re is calculated based on the inlet flow rate, provided that the piping parameters are established. Simulations are performed at various wall roughness levels to measure the pressure differential across the pipeline. The objective of this research is to enhance understanding of the characteristics of turbulent oil flow within straight pipelines. To this end, the simulations focus on the effect of varying wall roughness on the pressure differential at both ends of the pipeline, thereby gathering relevant data to assess pipeline linear loss and optimize pipeline design.

2.1. Physical Model

The length of the inlet section required to achieve fully developed flow is critical as it signifies the distance necessary to initiate turbulence within the conduit. In the context of a horizontal circular pipeline, this distance is represented by the length over which the fluid transitions from laminar to turbulent flow due to friction and other influencing factors. This length typically spans several times the diameter of the pipeline. The determination of this length takes into account factors such as pipeline geometry, fluid characteristics, and the Re. An accurate evaluation of the length of the inlet section in fully developed flow is essential for effective fluid system design, analysis, and a comprehensive understanding of fluid dynamics. The length of the inlet section, which marks the commencement of fully developed flow in a pipeline, correlates with the Re, as described by Equations (1) and (2) [27], which characterize turbulent states.

$$E_l = 4.4 \cdot \mathrm{Re}^{1/6} (\mathrm{Re} > 4000) \tag{1}$$

$$L = E_l \cdot d \tag{2}$$

In Equation (1), E_l represents the inlet section length factor, and L denotes the length of the inlet turbulence, while d indicates the pipeline diameter. Accompanying this description is a reference chart that illustrates the necessary inlet section length to achieve a fully developed flow. Figure 1 depicts a schematic of a fluid dynamics simulation for a horizontal circular pipeline, illustrating the concept of a fully developed flow and the turbulent inlet section. This schematic helps determine the minimum distance fluid must travel through the pipeline to reach the state referred to as "fully developed". In such a flow, the velocity profile remains constant along the pipeline axis, indicating no further changes in the flow dynamics.

The length of the inlet segment prior to the computational domain is crucial for the accuracy of the computations. An insufficient length in the entrance section leads to inadequate development of turbulence within the pipeline. In this study, the inlet section was measured to be approximately 0.7 m.

This study utilizes a quarter-circle pipeline model with uniformly distributed roughness to define the computational domain. Subsequently, the pressure differential between the two ends of the horizontally oriented pipeline is calculated. Comparative analyses between the complete circular pipeline model and the quarter-circle pipeline model consistently yield congruent results, indicating that the quarter-circle model effectively reduces the computational workload without compromising accuracy. This model is illustrated in Figure 2, where the pipeline is 1.2 m in length and has a radius of 20 mm. The model's length was chosen based on the 0.7 m length of the inlet section. In this study, a 0.5 m segment following the inlet section is selected as the subject of analysis to calculate the pressure difference across this portion of the pipeline. The spacing in the flow direction, denoted as *sf*, and the circumferential spacing, indicated as *sc*, are established. Furthermore, the height of the hemispherical roughness is represented as *h*. The fluid within the tube is specified as hydraulic oil with a density of 870 kg/m^3 and a dynamic viscosity of 0.04 kg/(m·s) at a temperature of 40 °C, as detailed in Table 1.

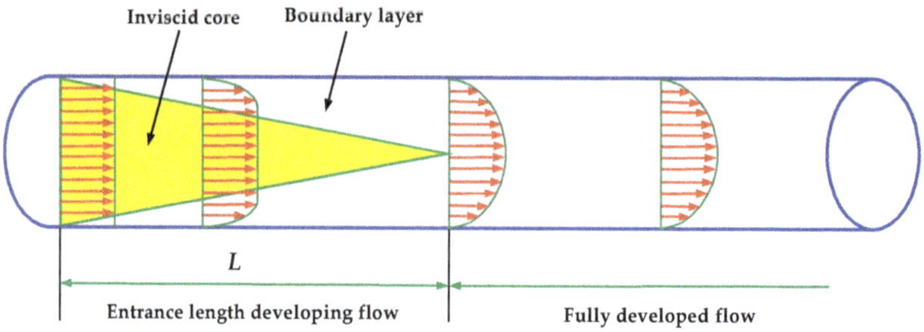

Figure 1. Attainment of full development of the inlet section length.

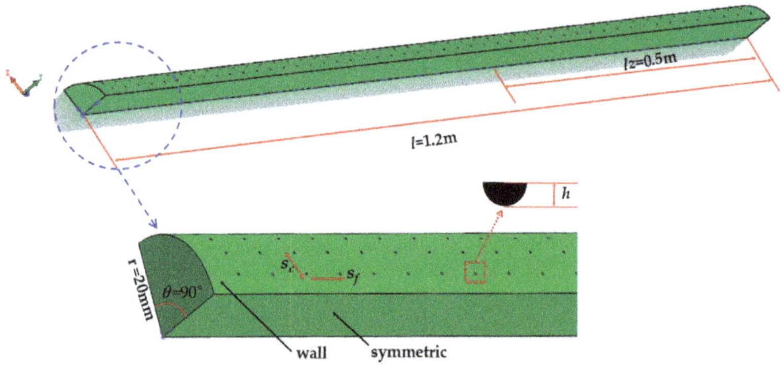

Figure 2. Quarter-circle pipeline calculation model with hemispherical roughness.

Table 1. Parameters of numerical simulation.

ρ (kg/m^3)	T (°C)	μ (kg/m·s)	v (m/s)	Re
870	40	0.04	6	5220

2.2. Mesh Generation

The resistance to fluid flow in a pipeline is predominantly caused by the interaction between the fluid and the pipeline wall, which generates a significant velocity gradient near the wall. To address this, the mesh near the wall, particularly around rough areas and along the wall-normal at their ends, is refined to ensure adequate resolution of these rough features. In this study, the flow regime, characterized by a low Re, requires a y+ value of approximately 1 to effectively resolve the boundary layer. The y+ is a dimensionless number that quantifies the distance from the wall where the fluid's viscous effects become significant. This method allows for a more precise capture of the near-wall flow dynamics. The near-wall boundary layer is modeled with at least eight mesh layers, employing a mesh growth rate of 1.2. The mesh growth rate describes the degree to which the sizes of adjacent meshes vary. This study rigorously validated grid convergence through detailed pressure and velocity analyses across various piping systems with grid densities ranging from 2 to 12 million cells. It utilized Richardson extrapolation in conjunction with Computational Geometric Information (CGI) techniques. The findings indicate a crucial observation: the discrepancy between predicted and actual pressure and velocity readings narrows to less than 1% at a grid size of 10 million cells. This suggests that enhancements in simulation accuracy become incremental beyond this grid density, confirming that a configuration of 10 million cells is sufficient for stable and accurate simulations of piping systems. Furthermore, CGI calculation outcomes support this conclusion, showing consistent pressure and velocity distributions across different grid densities, particularly at higher densities where variations stabilize. This underscores the effectiveness of using an 11 million grid size for precise hydrodynamic simulations of piping systems, optimizing computational resources. The mesh is divided as in Figure 3 and the hemispherical roughness elements are encrypted.

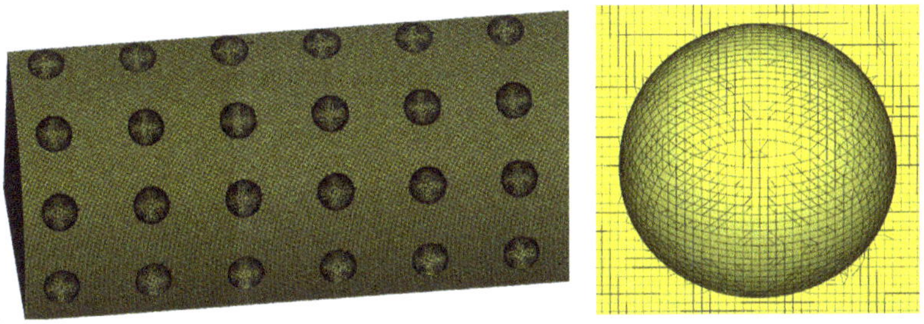

Figure 3. Rough cell meshing.

2.3. Boundary Conditions

The boundary conditions for the flow field simulation are meticulously defined to accurately reflect the state of the fluid at the beginning of the simulation. The inlet boundary condition specifies an inlet velocity of 6 m/s to precisely characterize the initial fluid dynamics. Conversely, the outlet boundary is established as a pressure boundary to effectively model the evolution of fluid pressure throughout the channel. The inner wall roughness of the pipeline is represented as a wall surface to faithfully simulate the interaction between the fluid and the rough surface. For turbulent flows, it is essential to allow turbulence to fully develop, ensuring uniformity in flow field characteristics both longitudinally and spanwise. This uniformity is critical for determining the accuracy of the simulation results, particularly in capturing the subtleties of flow in various directions. To minimize the effect of sidewalls on the flow field, the side boundaries are configured as symmetric. This setup prevents sidewall interference and facilitates a more realistic simulation of pipeline turbulence.

2.4. Turbulence Numerical Simulation

Investigating the flow fields over rough surfaces requires a detailed examination of the structure of the turbulent boundary layer. For this purpose, Reynolds-Averaged Navier-Stokes (RANS) simulations, which balance computational efficiency with the capability to capture micro flow field details, are particularly suitable. Therefore, RANS simulations are conducted in this study to analyze the flow field. Efficient computation is achieved by time-averaging the transient Navier-Stokes equations and utilizing the classical RNG turbulence model. The RNG k-ε model refines the computation of the turbulence length scale, which enhances the precision of turbulent viscosity estimations compared to the traditional k-ε model. This refinement is particularly effective in simulations of near-wall flows, where it accurately predicts complex turbulence phenomena adjacent to walls. Furthermore, the RNG k-ε model excels at forecasting flow separation phenomena, such as protrusions, by incorporating the impact of localized turbulence intensity variations on the flow. This aspect is crucial for examining wall-related studies and understanding the influence of geometric features on flow dynamics. Additionally, the augmented numerical stability and reliability of the RNG k-ε model ensure dependable outcomes in simulations involving complex boundary conditions and variable flow scenarios. Precise meshing is required to accurately delineate flow within the boundary layer; insufficient mesh refinement can lead to inaccuracies in predicting the flow field, especially near walls.

The Reynolds-averaged numerical simulation is performed for this calculation, with the governing equation detailed in Reference [28] expressed as follows:

Continuity Equation:

$$\frac{\partial \rho}{\partial t} + \frac{\partial}{\partial x_i}(\rho u_i) = 0 \tag{3}$$

Ds-averaged Navier-Stokes Equation:

$$\frac{\partial \overline{U}_i}{\partial t} + \overline{U}_j \frac{\partial \overline{U}_i}{\partial x_j} = -\frac{1}{\rho}\frac{\partial \overline{P}}{\partial x_i} + \frac{1}{\rho}\frac{\partial}{\partial x_j}\left(\mu \frac{\partial \overline{U}_i}{\partial x_j} - \rho \overline{u}_i \overline{u}_j\right) \tag{4}$$

The initial conditions are defined as [28]: The initial turbulence intensity is defined as:

$$I \approx 0.16\mathrm{Re}^{-\frac{1}{8}} \tag{5}$$

The initial turbulent kinetic energy is defined as:

$$k = \frac{3}{2}(\overline{U}I)^2 \tag{6}$$

The initial turbulence dissipation rate is defined as:

$$\varepsilon = C_{\mu}{}^{\frac{3}{4}} \frac{k^{\frac{3}{2}}}{l} \tag{7}$$

The $\bar{U}$ is mean velocity, C_μ = 0.09 is an empirical constant, l = 0.07L, and L is the characteristic length.

The discretization method employs a pressure-based SIMPLEC algorithm to enhance numerical stability by refining the pressure term in the discretized equations, thus mitigating the risk of numerical instability associated with the SIMPLE algorithm. Furthermore, a second-order upwind discretization scheme is utilized to improve accuracy and stability, offering significant advantages over the first-order scheme. Setting the convergence criterion at 10^{-5} ensures that the numerical simulation incrementally approaches the correct solution through iterative refinement, thereby avoiding excessive computation while meeting practical requirements [29].

3. Calculation of Different Roughness Parameters

The model incorporates hemispherical roughness arranged in a regular pattern on the wall. The selection of hemispherical roughness is informed by the findings from extensive domestic and international research cited in the introduction. This research indicates that hemispherical roughness is suitable for modeling the roughness of the inner walls of circular tubes, hence the adoption of this shape in the model. Key factors influencing fluid loss along the pipeline include the spacing and height of the roughness elements. By varying the spacing of these roughness elements, their impact on fluid flow within the pipeline can be quantified. We conducted separate analyses of the spacing and height of these roughness elements to clarify their effects on the inner wall surface and linear fluid loss. Consequently, the hemispherical roughness model addresses the localized effects of wall roughness on fluid dynamics, and adjusting these roughness characteristics enables the simulation of fluid loss along the pipeline.

3.1. Equal Spacing Goughness Calculation

In the near-wall region, wall conditions significantly influence flow velocity and distribution, highlighting the need for a detailed model to accurately represent fluid behavior due to pronounced interactions with roughness. Conversely, in the outer-wall region, the impact of roughness on flow is minimal, permitting the use of a simplified model to estimate fluid behavior further from the wall. Consequently, the model incorporating hemispherical roughness effectively accounts for the localized effects of wall roughness on fluid flow, facilitating the simulation of fluid loss along the pipeline by adjusting roughness characteristics.

The spacing of the roughness results in a slightly higher pressure differential at the pipeline end under irregular conditions compared to regular distribution conditions. This study contrasts irregular roughness distributions along the circumferential and flow directions, revealing that a regular distribution condition is more advantageous for minimizing losses along the pipeline. To further the research on the effects of a rough inner wall on fluid flow, we propose the introduction of another parameter: the spacing between roughness units in both the flow direction and the circumferential direction, denoted as sf and sc respectively. When sf equals sc, it indicates that the spacing of roughness units is uniform in both the longitudinal and circumferential directions. With the roughness height set to 0.03 mm and relative roughness at 0.00075, the flow in the pipeline is calculated for the dimensionless roughness spacings of $sf/\varepsilon = sc/\varepsilon$. Figure 4 illustrates the dimensionless roughness, and the corresponding values of differential pressure for equally spaced roughness.

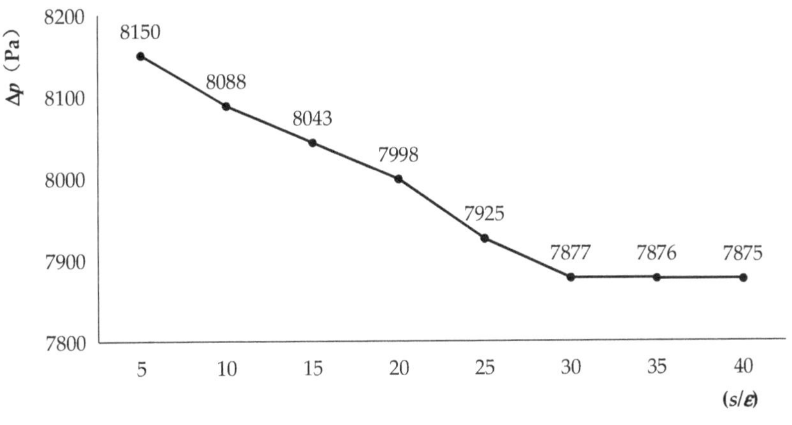

Figure 4. Differential pressure values for equidistant roughness.

The analysis of the calculation results depicted in Figure 4 demonstrates a nuanced relationship between surface roughness spacing and differential pressure. Specifically, an inverse relationship is observed for the dimensionless spacings less than 30: as the roughness spacing increases, the differential pressure decreases. Beyond this point, the rate of decrease in differential pressure slows, followed by a stabilization where the rate of pressure drop becomes very gradual and approaches a horizontal level. According to the Darcy-Weisbach equation (Equation (8)), the friction resistance coefficient is calculated as illustrated in Figure 5. As shown in Figure 5, the coefficient of frictional resistance along the pipeline begins to decrease gradually as roughness spacing increases. After the

the dimensionless spacing reaches 30, there are no further changes, indicating that the maximum value of the frictional resistance coefficient has been reached.

$$h_f = \lambda \frac{l}{d} \frac{v^2}{2g} \tag{8}$$

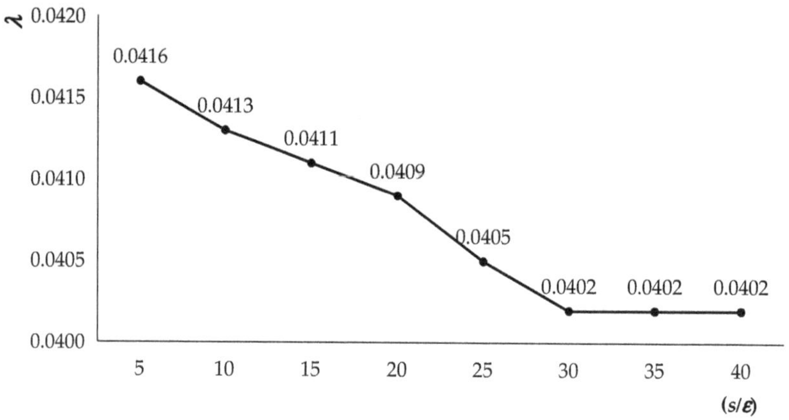

Figure 5. Friction resistance coefficient of different spacing.

Based on the roughness spacing determined in Figure 4, a consistent spacing of 30 was established. Differential pressure values between the two ends of the pipeline were calculated for varying roughness heights, with the results displayed in Figure 6. These roughness heights are based on the actual measurements taken from the hydraulic pipeline, hence the defined roughness heights. Observations from Figure 6 reveal that pressure loss in rough pipelines significantly exceeds that in smooth pipelines. Transitioning from the smooth to the rough tube, the rate of differential pressure increase significantly escalates. As roughness varies from 0.00025 to 0.00075, the rate of increase slows; however, further increasing the roughness causes the pressure growth rate to accelerate. As roughness increases from 0.00025 to 0.000125, there is a corresponding rise in differential pressure, consistent with the relationship described by the Darcy-Weisbach equation. The resistance coefficients for different roughness levels are shown in Table 2, illustrating that resistance coefficients increase with roughness.

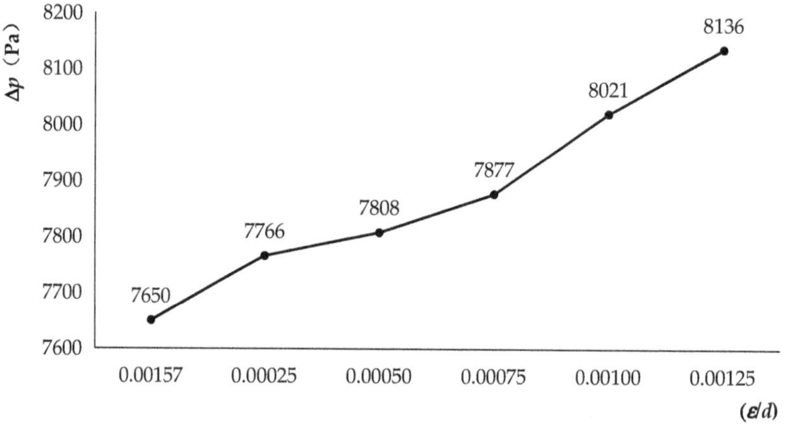

Figure 6. Differential pressure values for varied roughness height.

Table 2. Friction resistance coefficient of different roughness.

Frictional (λ)	Absolute Roughness (ε)	Relative Roughness (ε/d)
0.0391	smooth	-
0.0397	10 μm	0.00025
0.0399	20 μm	0.00050
0.0402	30 μm	0.00075
0.0401	40 μm	0.00100
0.0416	50 μm	0.00125

Figure 6 demonstrates that differential pressure in pipelines escalates with increased wall roughness, especially under turbulent flow conditions characterized by low Re. From a surface engineering perspective, applying nickel-metal coatings provides a solution by creating a smoother inner pipeline surface, thus diminishing the friction between the fluid and the wall. This smoothing effect is achieved by leveling out surface irregularities. The thickness of the coating plays a crucial role in reducing frictional losses; however, an excessively thick coating may restrict the pipeline's internal diameter, subsequently elevating flow resistance. Conversely, an overly thin coating may not sufficiently smooth out the rough patches on the inner wall. Therefore, identifying the optimal coating thickness is essential for ensuring sufficient smoothness without significantly reducing the pipeline's effective diameter. This investigation into drag reduction involved examining the effects of various coating thicknesses on longitudinal losses, with evaluations based on differential pressure variations at both pipeline ends. The findings from these numerical analyses are presented in Figure 6, with a baseline roughness of 0.00075.

Figure 7 illustrates the differential pressure values at both ends of a pipeline under various coating thicknesses. The data show that uncoated pipelines exhibit higher differential pressures, highlighting the efficacy of coatings in reducing these pressures. A marked decline in differential pressure is observed as coating thickness increases from 10 μm to 30 μm, indicating that thicker coatings improve the smoothness of the pipeline's internal surface, thereby reducing turbulence and associated pressure losses. The differential pressure value for pipelines with coatings is lower than that of uncoated pipelines under equivalent conditions of roughness height. However, at a coating thickness of 30 μm, the differential pressure in the pipeline exceeds that of a pipeline with a smooth inner surface. These findings emphasize the value of coatings as a strategic measure to enhance hydraulic efficiency and reduce operational costs in pipeline systems.

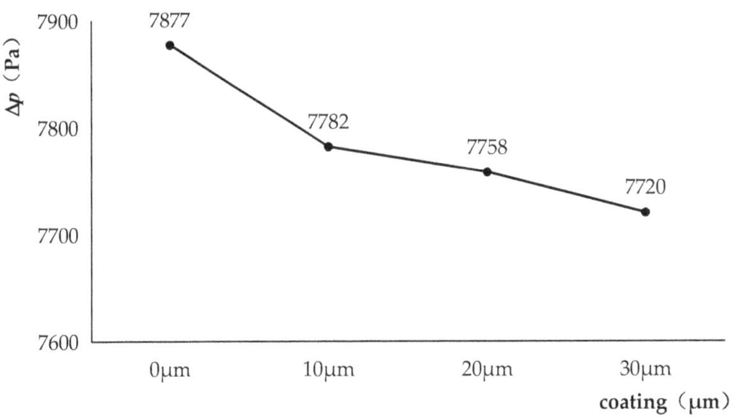

Figure 7. Pressure differential under different coating thickness.

3.2. Analysis of Drag Reduction Rate

The formula for calculating the drag reduction rate is detailed in Equation (9) [26]. According to this formula, there is a defined relationship between the loss incurred along the pipeline and the pressure difference. Consequently, this study employs the pressure difference to evaluate the drag reduction rate. Thus, the formula for the drag reduction rate can be equated to Equation (9):

$$\eta = \frac{\Delta h - \Delta h_{\text{coating}}}{\Delta h} \tag{9}$$

The evaluation of drag reduction facilitated by the coating involved analyzing changes in the pressure differential as defined in Equation (8). The drag reduction rate, indicative of the coating's effectiveness, was subsequently calculated. These results were compiled and illustrated in Figure 8, providing a clear reference for assessing and analyzing the drag reduction data in the context of this study.

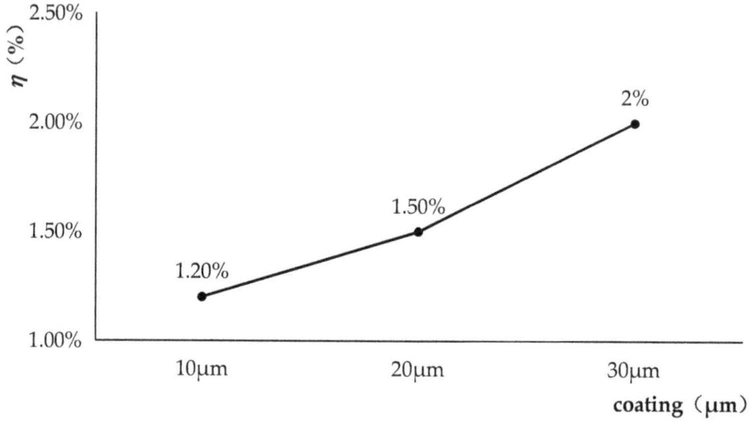

Figure 8. Drag reduction rate for different thicknesses of coatings.

Data analysis indicates that drag reduction decreases as coating thickness increases. A significant reduction of 2% was observed in samples with a 30 μm coating thickness. In contrast, samples with a 10 μm coating thickness exhibited the least reduction. Therefore, for practical applications where drag minimization is critical, maintaining a coating thickness of 30 μm is advisable.

4. Analysis of the Coating's Drag Reduction Mechanism

This study introduces a numerical simulation method which demonstrates that a coating thickness of 30 μm yields a notably smoother surface, significantly reducing frictional resistance within pipelines. The coating's effectiveness stems from its dual functionality: masking surface roughness and regulating the thickness of the viscous sublayer. The former directly reduces surface roughness, while the latter substantially affects the fluid's frictional resistance against the pipe wall. Thus, the coating not only diminishes roughness and modulates the viscous sublayer thickness but also enhances flow gradients and markedly decreases frictional resistance, reducing overall energy consumption.

The study explores the impact of coating thickness on the hydraulic performance of pipelines by evaluating various thicknesses. Specifically, coatings of 10 μm and 20 μm proved insufficient to conceal the 30 μm surface roughness of the pipeline wall. This inadequate coverage adversely affected the viscous sublayer, diminishing local flow velocity and increasing fluid friction. Conversely, a 30 μm coating effectively smoothes the rough surface of the pipeline and optimizes the flow path, thereby significantly reducing flow instability and turbulence and enhancing drag reduction. Nonetheless, the implications of utilizing excessively thick coatings were also considered. While a 30 μm coating adequately covers underlying roughness and improves overall smoothness, the material's intrinsic roughness may generate higher differential pressure than that observed in a completely smooth-lined pipe. This observation highlights the necessity to balance the coating's thickness and material properties to optimize pipeline performance while minimizing adverse hydraulic effects.

Velocity vector charts for both smooth and rough pipeline walls are presented in Figures 9 and 10. The disparity in fluid flow velocity gradients between soft and rough wall surfaces significantly influences hydraulic dynamics. Smooth wall surfaces exhibit a slight velocity gradient, indicating a gradual transition in velocity among fluid molecules, which enhances the stability of fluid flow. In contrast, rough wall surfaces display a more pronounced velocity gradient, suggesting abrupt changes in velocity among fluid molecules, thereby increasing susceptibility to turbulence and eddies, which contribute to unstable flow. Friction on smooth pipeline walls is minimal, leading to relatively low energy losses, whereas friction on rough walls is heightened, resulting in increased energy losses. A coating thickness of 30 μm effectively transforms rough wall surfaces into smoother ones.

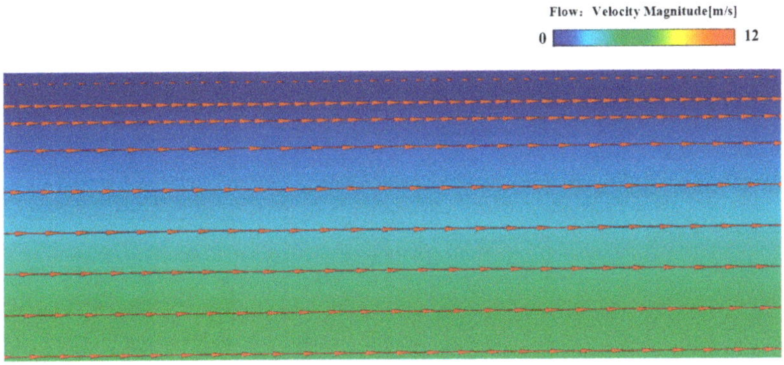

Figure 9. Smooth wall surface velocity.

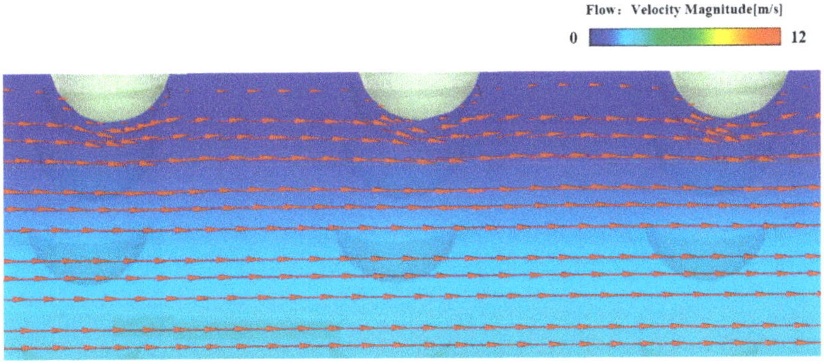

Figure 10. Rough wall surface velocity.

Figures 11 and 12 illustrate the surface roughness of velocity clouds before and after the application of a coating, respectively. The analysis reveals that the height of roughness decreases following the application of the coating to the rough wall surface, with the shape no longer retaining its initial hemispherical form. This modification lowers the fluid velocity gradient across the roughness surface, leading to smoother fluid flow within the pipeline. The reduced gradient facilitates a more uniform interaction among fluid molecules, thereby diminishing fluid fluctuations and eddies. Consequently, this smoother flow significantly lowers the differential pressure within the pipeline, enabling a more efficient passage of fluid. Thus, pipelines with coatings exhibit lower linear losses and reduced differential pressure values compared to their uncoated counterparts, given the same roughness height.

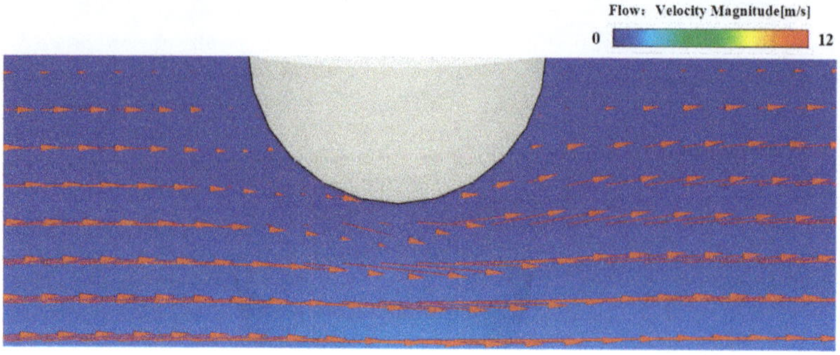

Figure 11. Uncoated rough surface velocity.

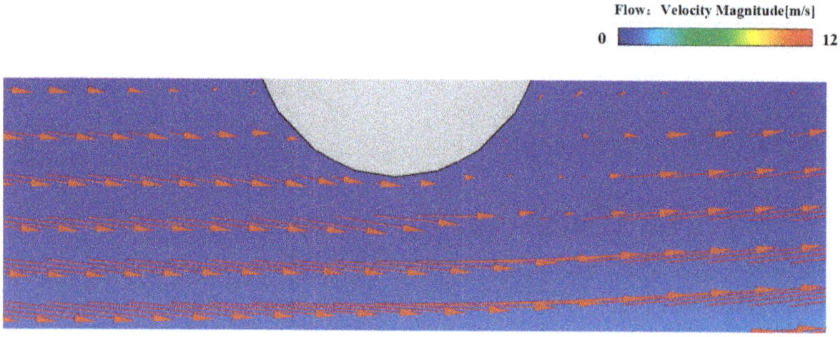

Figure 12. Rough surface velocity with coating.

5. Conclusions

This article examines the critical parameters influencing losses along turbulent circular pipelines, focusing on roughness spacing and coating thickness. It analyzes the effects of these parameters on turbulence and friction losses. The main findings include: Numerical simulations indicate that the characteristics of roughness, specifically the spacing, significantly impact fluid flow losses. By appropriately adjusting the roughness spacing, friction and turbulence losses can be substantially reduced. Stability in differential pressure values is achieved when roughness spacing is increased to 30. The study also investigates the effect of coating thickness on drag reduction. Results confirm that a 30 μm coating thickness optimizes drag reduction, achieving up to a 2% decrease. This thickness also contributes to a reduction in differential pressure for the same roughness height due to alterations in the roughness shape. Additionally, the 30 μm coating demonstrates higher linear losses compared to smooth pipelines, attributed to the intrinsic roughness of the coating itself. A detailed analysis of the drag reduction mechanism revealed that the dual role of the coating—covering the roughness and regulating the thickness of the viscous sublayer—was crucial. The 30 μm coating was found to be most effective in improving fluid transfer efficiency due to these factors. These findings underscore the importance of precise adjustments in roughness spacing and coating thickness for optimizing turbulence and drag reduction in turbulent circular pipelines.

This study primarily utilizes simulation calculations to assess the impact of surface roughness on loss within pipelines under turbulent flow conditions. To enhance the credibility and robustness of our findings, further research should validate the impact of surface roughness through experimental investigations. Future studies are recommended to integrate experimental data with simulation results, thereby enriching our understanding of fluid flow dynamics in pipelines with rough surfaces. This approach promises to offer a more comprehensive insight into the phenomena. Additionally, further exploration of coating techniques could generate innovative research concepts and develop more effective strategies for reducing pipeline drag. It is also advisable for subsequent studies to explore the effects at higher Re, building upon the foundational insights provided.

Author Contributions: Conceptualization and data curation, X.W.; data curation; methodology, W.L.; writing—review and editing, J.Z. and S.Y. All authors have read and agreed to the published version of the manuscript.

Funding: This research received funding from scientific research projects sponsored by various ministries and departments (Grant No. 3030021222316).

Data Availability Statement: Data from this study are available upon request from the corresponding author.

Acknowledgments: This research was carried out with the support and help of many people. First, we would like to express our deep gratitude to our project supervisors, Junjie Zhou and Shihua Yuan, whose expertise and valuable suggestions inspired this study. We are also very grateful to Wenbo Liao, who helped me immensely in experimental design and data analysis. In addition, we express our sincere gratitude to the Scientific research projects of ministries and commissions, whose financial support Grant No. 3030021222316 was essential for the conduct of this study.

Conflicts of Interest: The authors declare no conflicts of interest.

erences

Kim, J.H.; Lee, Y.M.; Lee, J.H.; Kim, J. Influence of the surface roughness on inner–outer interactions in a turbulent Couette–Poiseuille flow. *Phys. Fluids* **2021**, *33*, 045113. [CrossRef]
Wang, H.; Zhang, Y. Experimental study on the effect of wall roughness on turbulent drag reduction in a circular pipe. *J. Fluid Mech.* **2020**, *901*, A16.
Yu, X.; Ozdemir, C.E.; Yu, M.; Yu, Z. Properties of D- and K-type Roughness in Oscillatory Turbulent Boundary. *Environ. Fluid Mech.* **2021**, *21*, 885–905. [CrossRef]
Xu, P.; Zhao, W. Experimental analysis of drag reduction performance of different roughness configurations in pipe flows. *Exp. Fluids* **2019**, *60*, 24.
Heriyani, O.; Mugisidi, D.; Hilmi, I. Effect of cylinder surface roughness to the distance formation of vortex. *Sintek J.* **2020**, *14*, 94–98. [CrossRef]
Ashmawy, E.A. Effects of surface roughness on a couple stress fluid flow through corrugated tube. *Eur. J. Mech. B-Fluids* **2019**, *76*, 365–374. [CrossRef]
Han, Y.; Wang, S.Y.; Chen, J.; Yang, S.; Qiu, L.C.; Dharmasiri, N. Resistance of the flow over rough surfaces. *J. Hydrodyn.* **2021**, *33*, 593–601. [CrossRef]
Mirmanto, M.; Yudhyadi, I.; Sulistyowati, E.D. Effect of tube diameter and surface roughness on fluid flow friction factor. *Mech. Eng.* **2014**, *4*, 62–70. [CrossRef]
Heriyani, O.; Mugisidi, D.; Hilmi, I. Effect of the surface of the rough pipe on the fluid flow rate. *Mater. Sci. Eng.* **2020**, *909*, 012015. [CrossRef]
Du, D.; Li, Y. Numerical analysis of roughness effect on fluid flow in a micro tube with a three-dimensional roughness element model. In Proceedings of the 2009 International Conference on Engineering Computation, Hong Kong, China, 2–3 May 2009; pp. 182–185.
Yang, S.Q.; Han, Y.; Dharmasiri, N. Flow resistance over fixed roughness elements. *J. Hydraul. Res.* **2011**, *49*, 257–262. [CrossRef]
Song, S.; Yang, X.; Xin, F.; Lu, T.J. Modeling of surface roughness effects on stokes flow in circular pipes. *Phys. Fluids* **2018**, *30*, 023604. [CrossRef]
Wu, S.; Christensen, K.T.; Pantano, C. A study of wall shear stress in turbulent channel flow with hemispherical roughness. *J. Fluid Mech.* **2020**, *885*, A16. [CrossRef]
Mishra, A.V.; Bolotnov, I.A. DNS of turbulent flow with hemispherical wall roughness. *J. Turbul.* **2015**, *16*, 225–249. [CrossRef]
Klebanoff, P.S.; Cleveland, W.G.; Tidstrom, K.D. Tidstrom. On the evolution of a turbulent boundary layer induced by a three-dimensional roughness element. *J. Fluid Mech.* **1992**, *237*, 101–187. [CrossRef]

16. Zhou, Z.; Wang, Z.; Fan, J. Direct numerical simulation of the transitional boundary-layer flow induced by an isolated hemisph ical roughness element. *Comput. Methods Appl. Mech. Eng.* **2010**, *199*, 1573–1582. [CrossRef]
17. Guala, M.; Tomkins, C.D.; Christensen, K.T.; Adrian, R.J. Vortex organization in a turbulent boundary layer overlying spa roughness elements. *J. Hydraul. Res.* **2012**, *50*, 465–481. [CrossRef]
18. Iyer, P.S.; Mahesh, K. High-speed boundary-layer transition induced by a discrete roughness element. *J. Fluid Mech.* **2013**, *7* 524–562. [CrossRef]
19. Chatzikyriakou, D.; Buongiorno, J.; Caviezel, D.; Lakehal, D. DNS and LES of turbulent flow in a closed channel featurin pattern of hemispherical roughness elements. *Int. J. Heat Fluid Flow* **2015**, *53*, 29–43. [CrossRef]
20. Qin, Y.; Zhao, J.; Liu, Z.; Wang, C.; Zhang, H. Study on effect of different surface roughness on nanofluid flow in nanochannel using molecular dynamics simulation. *J. Mol. Liq.* **2021**, *346*, 117148. [CrossRef]
21. du Toit, C.G. Fundamental evaluation of the effect of pipe diameter, loop length and local losses on steady-state single-pha natural circulation in square loops using the 1D network code Flownex. *Therm. Sci. Eng. Prog.* **2021**, *22*, 100840. [CrossRef]
22. Garcia, A.; Morini, G.L. Effect of roughness elements on drag reduction in turbulent pipe flows. *J. Fluids Eng.* **2020**, *142*, 0813
23. Zhu, T.; Wang, Y.; Chen, J. Effect of roughness shape on turbulent drag reduction in circular pipes. *Exp. Therm. Fluid Sci.* **20** *131*, 110528.
24. Gu, Y.Q.; Fan, T.X.; Mou, J.G.; Wu, D.H.; Zheng, S.H.; Wang, E. Characteristics and mechanism investigation on drag reduction oblique riblets. *J. Cent. South. Univ.* **2017**, *24*, 1379–1386. [CrossRef]
25. Smith, D.J.; Jones, M.R. Optimization of roughness pattern for turbulent drag reduction in pipelines. *Flow Turbul. Combust.* **20** *106*, 345–358.
26. Wang, X.; Zhou, J.; Yao, B.; Liao, W. Analyzing the efficacy of nickel plating coating in hydraulic pipeline drag reducti *Lubricants* **2024**, *12*, 37. [CrossRef]
27. Frank, M. White. In *Fluid Mechanics*; McGraw-Hill Education: New York, NY, USA, 2010.
28. Dellavalle, R. *Turbulence Modeling for CFD*; DCW Industries: La Canada, CA, USA, 2006.
29. Matos, D.; Valerio, C. *Fluid Mechanics and Pipe Flow: Turbulence, Simulation and Dynamics*; Nova Science Pub Inc.: Hauppauge, N USA, 2009.

coatings

icle

esearch on the Measurement and Estimation Method of Wheel esistance on a Soil Runway

an Wang, Xiaolei Chong, Guanhu Wang *, Chaojia Liu and Jichao Zhang

College of Aeronautical Engineering, Air Force Engineering University, Xi'an 710038, China
* Correspondence: wangguanhu@126.com

Abstract: In view of the fact that there is no suitable measurement method for tire driving resistance during the operation of a transporter on soil pavement, this paper proposes a simple measurement and estimation method for driving resistance on a soil runway based on the Bekker settlement model. In this paper, using the driving resistance equation based on the Bekker model, a simplified equation of driving resistance related to mass is proposed. Using the principle of kinematics, a simple test device for driving resistance was developed, and the test data of the test device were used to determine the parameters of the driving resistance estimation equation of the road surface. The driving resistance of the wheel running state was tested by the simulated aircraft load loading vehicle through the mechanical sensor, and the consistency between the theoretical value and the actual value of the driving resistance value under 8T, 10T, and 12T loads was verified. This research shows that the test method and estimation method have low cost and good accuracy and are suitable for wide promotion. The results of this paper provide a new idea and method for estimating the driving resistance of a wheel when a transporter runs on a soil road and also provide a reference for designing the length of a soil runway.

Keywords: soil pavement; wheel driving resistance; test method; estimation method

1. Introduction

The capability of modern transport aircraft to take off and land on unpaved runways represents one of its vital performance indicators. The distance required for these operations on such runways is intimately tied to the resistance encountered by the aircraft on the surface, which is a crucial factor. A relatively comprehensive system has been established for measuring resistance during takeoff and landing on concrete and asphalt pavements. On these surfaces, the primary source of resistance stems from friction; hence, the study of pavement resistance is commonly referred to as skid resistance research, defined as the investigation of friction within the contact zone between tires and the pavement surface [1]. The exploration of pavement skid resistance originated from the broader study of pavement performance, which gained momentum internationally in the 1930s. Early research on skid resistance was primarily initiated in countries experiencing abundant rainfall or frequent snow and ice, where slipping hazards posed significant challenges, thereby necessitating such investigations. The complexity of skid resistance research arises from the myriad of factors influencing the anti-slip performance between tires and road surfaces. From the perspective of the contact system, the primary influences stem from tire properties, pavement characteristics, and surrounding environmental conditions [2]. Specifically, the skid resistance of both concrete and asphalt pavements comprises the following essential components: adhesion and hysteresis [1]. These two components are intimately linked to the micro- and macro-textures of the pavement surface, emphasizing the pivotal role of pavement texture in determining skid resistance. Thus, the study of skid resistance necessitates a comprehensive understanding of these texture characteristics and their interplay with the various factors influencing tire–pavement interaction.

In the past few decades, experts and scholars at home and abroad have proposed many different types of detection equipment and methods. These methods can be divided into direct measurement methods and indirect measurement methods according to the category [3]. The direct approach to assessing pavement skid resistance involves utilizing index parameters that focus on measuring the friction coefficient, a core metric. Conversely, the indirect method revolves around evaluating pavement characteristics and inferring skid resistance based on established conversion relationships, with a primary emphasis on pavement texture analysis. Both methodologies continue to advance through innovation and refinement. One of the earliest methods for testing pavement skid resistance was the vehicle braking distance measurement [4,5]. This straightforward technique involves conducting braking tests on wet roads at 64 km/h, measuring the distance traversed until the wheels lock and the vehicle comes to a halt. It boasted versatility as it could be conducted using regular vehicles, yet its practicality was hindered by limitations in the supporting equipment at the time, leading to a cumbersome process and potential safety concerns. The British Pendulum Test (BPT) [6], introduced by the British Road Research Laboratory in 1964, is a cost-effective and user-friendly tool for evaluating pavement surface friction characteristics. Widely adopted globally, it excels in assessing pavement apparent micro-texture. However, its application is constrained by its inability to cover large areas or perform continuous measurements efficiently. Complementing the pendulum friction tester, the Dynamic Friction Tester (DFT) [7] offers a versatile alternative for low-speed skid resistance assessments. With an extended speed range and the capability for continuous friction monitoring, DFT boasts portability, high efficiency, and swift data processing. Its outcomes can be translated into the International Friction Index (IFI), facilitating standardized comparisons. The Locked-Wheel Skid Trailer [8], prevalent in the United States for both concrete and asphalt pavement evaluations, conducts friction tests on wet surfaces under constant load and a velocity of 64 km/h. It distinguishes itself with its ability to provide continuous friction measurements without disrupting road usage, eliminating the need for lane closures. The Grip Tester [9], a fixed-slip device, specializes in assessing the friction of smooth tires at slip rates ranging from 12% to 20%. Although it delivers precise measurements, its implementation is costly, requiring extensive technician support and significant investment. Lastly, the Mu-Meter [10] represents a versatile lateral force testing solution, capable of assessing a wide array of road conditions, including straight, curved, and steep sections. This equipment underscores the evolution of skid resistance measurement techniques towards more comprehensive and adaptable solutions.

Presently, research on the aircraft running resistance method on soil runways is mainly limited to a series of studies carried out by the United States military. In order to build soil runways for C-17, C-130, and other transport aircraft, the US military used high-speed continuous friction measurement equipment to test the resistance of simple runways. Tingle [11] used a continuous friction test device based on a trailer to perform continuous friction measurements along the direction parallel to a runway in order to develop a soil runway condition level that supports the operation of the C-17 transporter. Ward [12,13] proposed a friction test method for unpaved runway surfaces based on deceleration, which assists in the development of friction prediction models on roads with different humidity conditions and soil types. Based on the above, it can be concluded that many types of detection equipment for concrete and asphalt pavement have been gradually extended to soil pavement. Although there are few case studies on the running resistance of aircraft wheels on soil road surfaces in the field of airport engineering at home and abroad, the research on the interaction between wheels and soil in the field of agricultural and industrial machinery started earlier [14]. Mcallister [15] used a trailer to measure the rolling resistance coefficient of agricultural machinery tires and analyze the influencing factors. Their results showed that tire pressure, vertical load, tire size, and other factors have a great influence on the rolling friction coefficient. Rebati [16] compared the rolling resistance data of different tires with the calculated values of various models and evaluated the prediction accuracy of rolling resistance. They found that the EMOB model was better at predicting cross-layer tires and there was no better model for radial tires. Elwaleed [17] studied the influence of tire pressure on driving resistance and used three different tire pressures to test resistance. The test results showed that the logarithmic model was the best way to describe the movement resistance of agricultural tires. Botta [18] conducted field tests under three soil types, two tire types, and axle load conditions to confirm the relationship between the existing ground pressure parameters and tire shrinkage. The research in the related fields mentioned above studies the resistance of tires on soil road surfaces from various perspectives such as test methods, prediction models, and influencing factors. However, because of the complexity of soil properties, the research conclusions are not universal, and the process is more complicated.

Using a mechanical model of the interaction between aircraft tires and soil pavement based on Bekker theory and its motion equation, this paper explores the driving resistance estimation method of a wheel on a soil runway. The estimation method consists of two important parts. One is to measure the driving resistance of the vehicle based on the kinematics principle, and the other is to estimate the driving resistance of the large-weight aircraft based on the theoretical equation of the driving resistance and the measured driving resistance. This research provides a simple method for estimating tire resistance when an aircraft runs on a soil runway.

2. Theory and Methodology

2.1. Characterization Method of Wheel Resistance on Soil Road Surfaces

2.1.1. Driving Resistance Characterization Method Based on the Mechanical Model

When a wheel runs on a soil surface, the interaction model between the wheel and soil assumed by Bekker, an American scientist, in the 1960s is usually adopted for analysis [19], and its principle is shown in Figure 1.

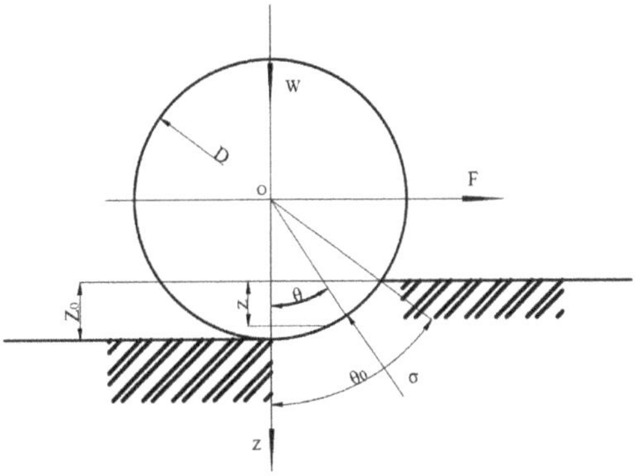

Figure 1. Simplified tire-soil interaction model.

In the figure, D is the outer diameter of the wheel, W is the vertical load, F is the driving resistance, Z_0 is the maximum subsidence depth of the tire, z is the subsidence depth, σ is the radial pressure, θ is the angle between the radial pressure and the vertical direction, and θ_0 is the penetration angle of the tire.

For dirt surfaces, pneumatic tires are similar to rigid wheels. Bekker assumed that the reaction force at all points of the surface on the contact area was radial and equal to the normal pressure under the same depth plate in the pressure-subsidence test. According to this assumption, the simplified action model below is drawn to obtain the balance equation of the driven rigid wheel, as shown in Equations (1) and (2):

$$F = b \int_0^{\theta_0} \sigma r \sin\theta d\theta \quad (1)$$

$$W = b\int_0^{\theta_0} \sigma r \cos\theta d\theta \tag{2}$$

where F is the driving resistance, b is the tire width, r is the tire radius, σ is the normal pressure, and W is the vertical load.

At the same time, the model satisfies Bekker's pressure–settlement relationship equation, as shown in Equation (3):

$$p = \left(\frac{k_C}{b} + k_\varphi\right) z^n \tag{3}$$

where p is the pressure, b is the tire width, and n, k_C and k_φ are the parameters describing the pressure–subsidence relationship.

By combining the above relationships and geometric configuration, the driving resistance equation of tire on soil pavement is established, as shown in Equation (4). For the specific derivation process, see reference [20]:

$$F = \frac{1}{(3-n)^{\frac{2n+2}{2n+1}}(n+1)b^{\frac{1}{(2n+1)}} \cdot (k_C/b + k_\varphi)^{\frac{1}{2n+1}}} \left(\frac{3W}{\sqrt{D}}\right)^{\frac{2n+2}{2n+1}} \tag{4}$$

Assuming Equation (5):

$$A = \frac{\left(\frac{3g}{\sqrt{D}}\right)^{\frac{2n+2}{2n+1}}}{(3-n)^{\frac{2n+2}{2n+1}}(n+1)b^{\frac{1}{(2n+1)}} \cdot (k_C/b + k_\varphi)^{\frac{1}{2n+1}}} \tag{5}$$

then, the driving resistance can be characterized as Equation (6):

$$F = Am^{\frac{2n+2}{2n+1}} \tag{6}$$

In Equation (6), n can confirm the value based on consulting relevant data. A is the comprehensive coefficient affected by factors such as soil characteristics and tire conditions. When the test vehicle and soil conditions are the same, A is a fixed value.

2.1.2. Kinematic Model-Based Driving Resistance Characterization Method

When an aircraft runs on a soil surface, the takeoff and landing processes are typical acceleration and deceleration processes, and the kinematic equation is as shown in Equation (7):

$$S = v^2/2a \tag{7}$$

where v is the landing speed or departure speed, S is the landing or takeoff distance, and A is the acceleration. In actual aircraft operation, whether takeoff or landing, this is a variable acceleration process, which is often simplified to a uniform acceleration or deceleration process in the calculation process.

Taking the landing process of an aircraft as an example, according to Newton's second law, the resistance $F = ma$ encountered by an aircraft during operation is composed of two parts, road surface resistance and air resistance. This paper mainly studies the resistance from the road surface, that is, the wheel driving resistance. From the perspective of kinematics, the resistance can be characterized as Equation (8):

$$F = mv^2/2S \tag{8}$$

2.2. *Field Testing of Driving Resistance Based on the Mechanical Model*

Bekker's classic model is an empirical equation, which is mainly applicable to the scenario of inflatable tires on soil surfaces. In order to verify the accuracy of its application in the calculation of aircraft drag on soil runways, a specific site is selected, a soil runway is built, and a simulated aircraft loading vehicle is used to carry out field tests.

2.2.1. Basic Conditions of the Test Site

The test site is located in Baqiao District, Xi'an, Shaanxi Province, with an altitude of about 420 m, and the soil is sticky soil. The Xi'an area has a temperate subhumid continental monsoon climate and is hot and rainy in summer, thundery and windy weather, cold and dry in winter. The test time was from early June to early July, during which the weather is hot, the average temperature is 31 °C, the highest temperature is more than 40 °C, and the precipitation is less, which is suitable for field tests. The test section site is shown in Figure 2.

Figure 2. Test section field.

The depth treatment of the structural layer of the pavement is 30 cm. Because there is no bad geological condition in the original foundation, there is no need to treat the original foundation. Humus is cleaned during construction, and debris such as stones and plastics also need to be removed. The soil in this test section is of good quality and is directly compacted after removing impurities. In terms of the compaction process, the compaction degree reaches more than 90% after a 30T heavy roller is used to vibrate the compaction surface 6 times.

2.2.2. Test Method

(1) Test equipment

Simulate loading vehicle equipment.

Traffic tests were carried out using a simulated aircraft loading vehicle [21]. The aircraft simulation loading vehicle can simulate the aircraft load, with a self-weight of about 70 kN, equipped with 10 kN weight blocks, up to 200 kN. In the loading process, it can realize step-by-step loading. Figure 3a shows the simulation aircraft load-loading vehicle.

(**a**) Loading vehicle appearance (**b**) Sensor on the wheel

Figure 3. Aircraft load simulation loading vehicle.

The tension sensor is installed on the bearing wheel, and the heading force is tested by the sensor. The sensor is installed on the central air tire (see Figure 3b) to monitor the direction and force of the tire in real time.

(2) Test process.

Starting from one end of the test section, a tractor is used to make the aircraft load simulate the loading vehicle pass through the test section at a constant speed at a speed of 10 km/h. The weight of the loading vehicle is set to 8T, 10T and 12T respectively, and the same weight is tested 10 times. During the test, the sampling frequency is 10 Hz, and the force in the forward direction is obtained directly through the sensor. The site diagram of the aircraft load simulation loading vehicle is shown in Figure 4.

Figure 4. Loading truck field test.

In this test, it was found that the rut depth did not change significantly after 10 tests, so the test times were set to 10 times.

2.3. *Field Test of Driving Resistance Based on the Kinematic Model*

According to the kinematics equation, as long as the mass, speed, and distance of the object in the process of movement are measured, acceleration can be inversely calculated, and then the driving resistance can be obtained. According to this principle, the test scheme was designed as follows.

2.3.1. The Test Principle of Speed and Distance

By determining the change rule of the position of the object with time, the running speed and distance can be calculated. At present, the most commonly used position measurement tool is the satellite positioning system. During the test, the satellite positioning system was used, as shown in Figure 5 The model was GPS320 (see Figure 5a), which supports the full-band signals of GNSS's four systems (GPS, GLONASS, BDS, GALILEO). The antenna positioning is accurate to the millimeter level. The lower part of the antenna is equipped with a magnetic chassis, so that the antenna can be stably adsorbed on the object during the test. The main body of the antenna is stably connected to the GPS receiver through the TNC line (see Figure 5b,c).

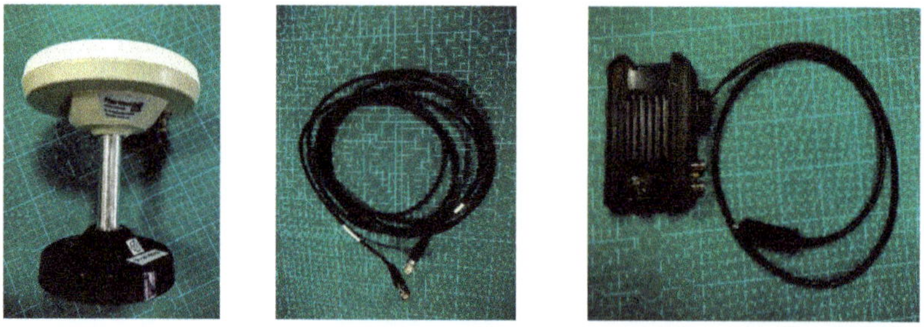

(**a**) (**b**) (**c**)

Figure 5. Position measurement equipment. (**a**) GNSS antenna; (**b**) TNC line; (**c**) GPS receiver.

Through this device, the vehicle speed information received by the GNSS instrument is transmitted to the mobile phone APP, and the image and CSV data are output and presented on the mobile phone APP (see Figure 6). The APP was developed based on the Android operating system and can be run on smartphones. This APP was developed using the Android Studio development tool. The test state is changed by the change curve of speed and time, and the required data such as the deceleration speed and deceleration distance are counted. The interface is shown below. When the test goes through the start, acceleration, stationary, deceleration, and end states, complete data information is formed. This data information includes start time, start deceleration time, end deceleration time, start deceleration speed, total deceleration time, deceleration distance, test time, test state, deceleration speed curve, deceleration distance curve, etc. It also outputs images of deceleration time and deceleration speed and deceleration time and deceleration distance. Digital data information can be output in CSV data format.

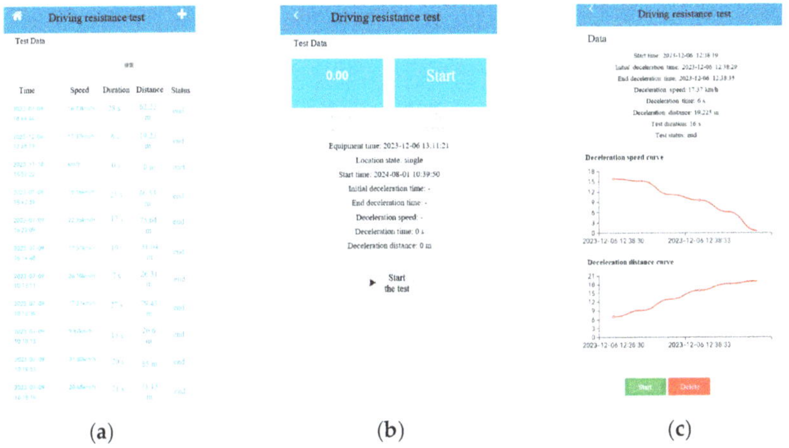

Figure 6. Wheel resistance test APP page. (**a**) Test record Page; (**b**) Test operation page; (**c**) Test data Page.

2.3.2. Test Vehicle Selection

This test used the common pickup truck in China as the test vehicle, and the weight of the vehicle is about 2T. The GNSS antenna is installed on the roof of the vehicle, and the signal processor is connected to the power supply in the vehicle to receive the signal. The tire model of the vehicle is 235/75R15, and the tire pressure is 35 psi. The size of the vehicle is shown in Figure 7.

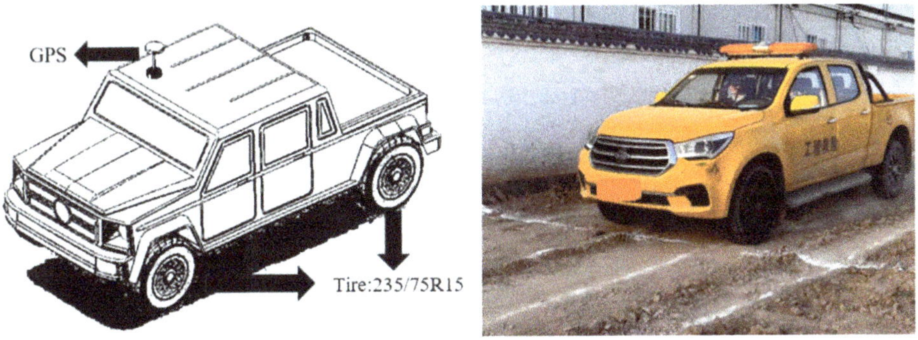

Figure 7. Vehicle size.

2.3.3. Test Procedure

In the soil field test section established in the second part, the vehicle is set to accelerate from static until the speed reached the predetermined speed, and then the throttle is loosened to allow the vehicle to slide freely forward in a straight line, and finally the speed is reduced to 0 without braking. The vehicle starts from the initial deceleration speed on the test section until the speed drops to 0, which is to complete a complete test. Considering the dispersion of deceleration data and the possible inaccuracy in the device, at least ten repeated measurements were performed for each scene.

3. Results and Discussion

3.1. Test Result Analysis of the Mechanical Model

After the outlier processing of the resistance data obtained from the aircraft load simulation vehicle test, the curve of resistance with time and a box diagram were drawn, as shown in Figure 8. In addition, the standard deviation of each group of data is shown in Table 1.

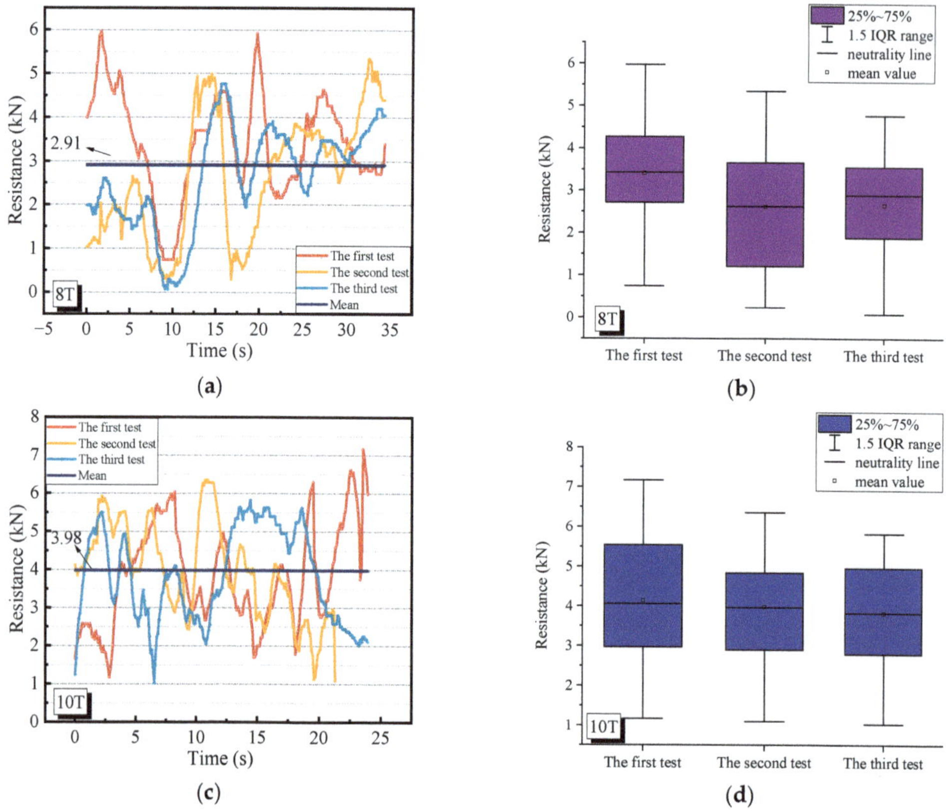

Figure 8. *Cont.*

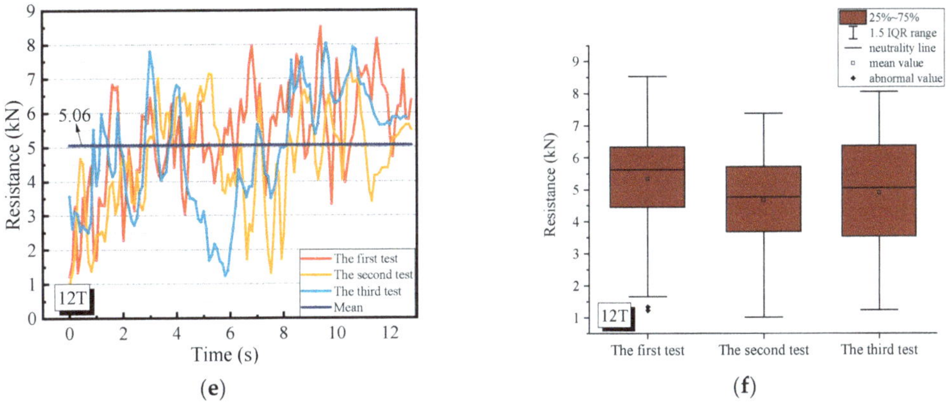

Figure 8. Driving resistance value change diagrams for 8T, 10T, and 12T. (**a**) Resistance value line chart, 8T; (**b**) Resistance value box diagram, 8T; (**c**) Resistance value line chart, 10T; (**d**) Resistance value box diagram, 10T; (**e**) Resistance value line chart, 12T; (**f**) Resistance value box diagram, 12T.

Table 1. Driving resistance variance.

	8T			**10T**			**12T**		
σ	1.16	1.65	1.18	1.42	1.23	1.21	1.51	1.52	1.71
μ	3.39	2.69	2.65	4.14	3.97	3.82	5.35	4.67	4.89

The figure shows that the image drawn from the data collected at a sampling frequency of 10 Hz has good continuity. On the soil road surface with a 90% compaction degree, the variation law of road surface resistance under different loading vehicle weight conditions is similar, which fluctuates up and down around a certain value, and the fluctuation range is large, which is also reflected in the box diagram and standard deviation. The variation range of pavement resistance under the three loading conditions is basically between 0 and 6 kN, and most of them are concentrated in the range of 2~5 kN. In addition, with the increase in load, the concentration range of pavement resistance increases gradually, the fluctuation range also increases obviously, and the concentration of data is better. The figure intuitively shows that when the weight of the loading vehicle is 12T, the maximum value of the resistance value exceeds 8 kN, which is higher than the range of the 8T and 10T loads. The average value of the above data was calculated. The average value of the pavement resistance under the weights of the 8T, 10T and 12T loading vehicles shows an upward trend, which is 2910 N, 3978 N, and 5055 N, respectively. In general, although the internal stability of each set of data is poor, the mean value of each set of measured data is relatively stable, and the tested driving resistance is more accurate.

3.2. Test Data Analysis of the Kinematic Model

The data output by the test processing software (Wheel resistance test 1.0) mainly includes the initial deceleration speed, deceleration distance, deceleration time, and so on. The output curve is the curve image of deceleration time and deceleration speed and deceleration time and deceleration distance, as shown in Figure 9.

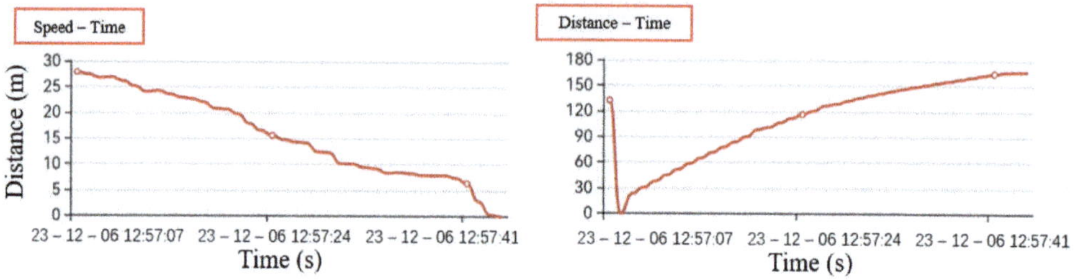

Figure 9. Deceleration speed curve and deceleration distance curve.

The test data were recorded at a frequency of 2 Hz, that is, twice per second. In order to facilitate the calculation, the acceleration reduction in the vehicle in the sampling interval remained unchanged, which is a uniform deceleration process. On this basis, the equation is introduced as Equation (9):

$$a = \frac{v^2 - {v_0}^2}{2S} \tag{9}$$

where a represents the deceleration between the sampling intervals, S represents the distance between the sampling intervals, v represents the speed before the sampling interval, and v_0 represents the speed after the sampling interval.

The deceleration acceleration between the sampling intervals was calculated, and the resistance was obtained by combining Newton's second law. A total of 10 experiments were carried out in this experiment, and 10 sets of effective data were obtained. In order to reflect the data characteristics more clearly, four sets of data were randomly selected for image analysis. The change in resistance with time is shown in Figure 10, and the resistance box diagram and scatter plot are shown in Figure 11.

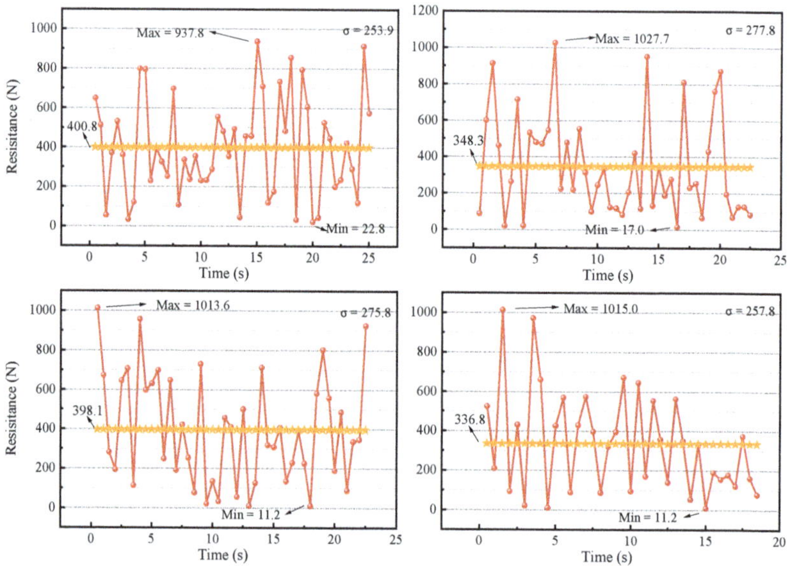

Figure 10. Resistance value change diagram.

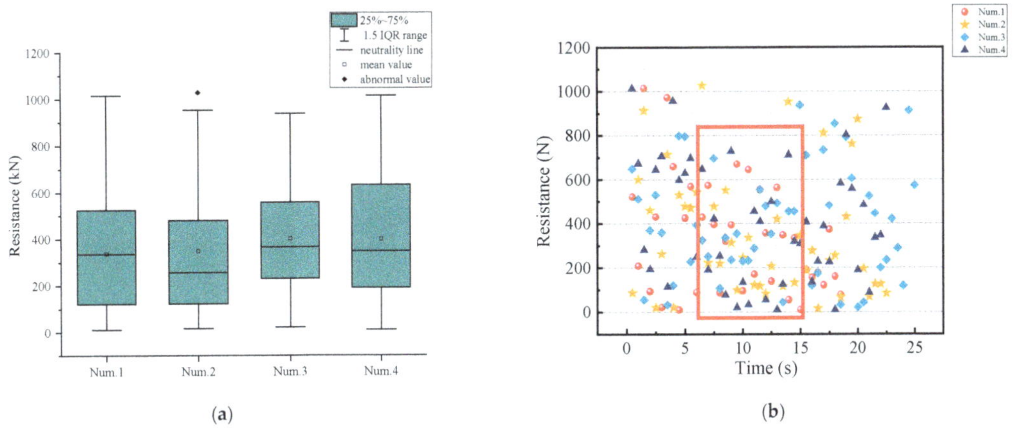

Figure 11. Resistance value box diagram and scatter plot. (**a**) Resistance value box diagram; (**b**) Resistance value scatter plot.

The diagram shows that the size of the resistance goes up and down, the overall trend in the fluctuation is stable, and there is no upward trend or downward trend. Because of the difference in the soil compaction degree, the change in soil flatness, the change in the contact condition between tire and soil surface, etc., the resistance of the vehicle will change in the process of moving forward. From the mean value of multiple tests, the mean value of each test dataset is almost equal, and the change in resistance fluctuates up and down around the mean value. Therefore, this paper assumes that this mean value can be used as the resistance value of this test. As shown in the box diagram, the variation range of resistance is 0~1000 N, and most of the data are concentrated between 200 and 700 N. From the time point of view, Figure 11 accurately reflects the characteristics of the change in resistance value. At the beginning and end of the test, the data are more discrete, while the data in the middle (see the red box in Figure 11b) of the test are more concentrated and less volatile. Through the measurement and analysis of multiple sets of data, the average value of multiple measurement data was taken, and the resistance on the compacted soil road surface of the vehicle is 381 N. Although the internal fluctuation of each set of data is large, its mean is relatively stable. The difference between the mean value of each group of data and the mean value of the four groups of data is within 10%, which can reflect the reliability of the measured value.

3.3. Estimation Method of Aircraft Running Resistance on the Soil Road Surface

3.3.1. Determination of the Calculation Parameters of Road Surface Driving Resistance

1. Principle of parameter determination.

Before using Equation (6) to calculate the driving resistance, two parameters A and n need to be determined. A is determined by the soil characteristics and tires. Relevant scholars [20] have carried out a large number of tests on uniform soil and given the recommended values of parameters such as n, k_C, and $k\varphi$ under typical soil conditions. According to the recommended values and tire characteristics, A of the corresponding soil and tire can be calculated, and n can be determined according to the type of soil. The results of relevant scholars show that Equation (6) is reasonable in calculating the driving resistance of tires. However, in the actual use process, the soil environment is complex, and it is difficult to be consistent with the working conditions of the recommended parameters. Therefore, it is very important to determine the accurate parameters for the calculation of the driving resistance of aircraft tires on a soil runway.

According to reference [20], n can be determined according to the type of soil, then A can be characterized as Equation (10):

$$A = \frac{F}{m^{\frac{2n+2}{2n+1}}} \tag{10}$$

Corresponding to a certain soil pavement, as long as F under a certain working condition is tested, the corresponding coefficient A can be calculated. Under the action of the tire, the resistance of the tire on the soil with a different compaction degree of the same soil can be calculated, which avoids the error caused by the variation in soil parameters and simplifies the calculation method.

2. Determination of calculation parameters based on the actual test.

Aiming to calculate the driving resistance of the soil road surface, based on the principle of kinematic resistance characterization, the test method established in the fourth part was used to carry out the field resistance test. In this paper, the soil of the test site is clay. According to the literature [20], the n value is 0.5. Under this soil condition, when the compaction degree is 90%, the driving resistance of the corresponding 2T vehicle is 381 N, and the A value is 134.7. Therefore, under the condition of the soil pavement, the calculation equation of the resistance under different vehicle weights without changing the test tire parameters is as shown in Equation (11):

$$F = 134.7m^{1.5} \tag{11}$$

Corresponding to the given soil road, especially the temporary soil simple road, the vehicle weight can also be estimated by this equation. If the resistance caused by the weight is greater than the traction provided by the engine, the vehicle cannot pass.

3.3.2. Estimation of Road Surface Driving Resistance

The idea of estimating the driving resistance of the road surface is as follows: for the same soil runway, according to the similarity principle, based on the kinematic equation, the resistance of the vehicle on the road surface is tested by vehicle motion, and then, the vehicle driving resistance is used to estimate the aircraft driving resistance. This method will produce a certain error. The error comes from the difference between the size of the aircraft tire and the size of the test vehicle. Different tires will cause differences in coefficient A. Ignoring the influence of this error, this idea was applied to estimate the driving resistance of the road surface. Taking the data in this paper as an example, considering the single wheel loads of 8T, 10T and 12T, the corresponding wheel driving resistance is calculated as 3048 N, 4260 N, and 5600 N by applying the equation $F = 134.7m^{1.5}$.

3.3.3. Accuracy Analysis of the Estimation Method

In the second part of this paper, the pavement resistance under different single wheel loads was tested on site. The comparison between the test data and the estimated value is shown in the Table 2 and Figure 12.

Table 2. Data comparison analysis table.

$y = 134.7x^{1.5}$			
Mass	**Theoretical Value**	**Actual Value**	**Degree of Deviation**
8T	3048	2910	4.53%
10T	4260	3978	6.62%
12T	5600	5055	9.73%

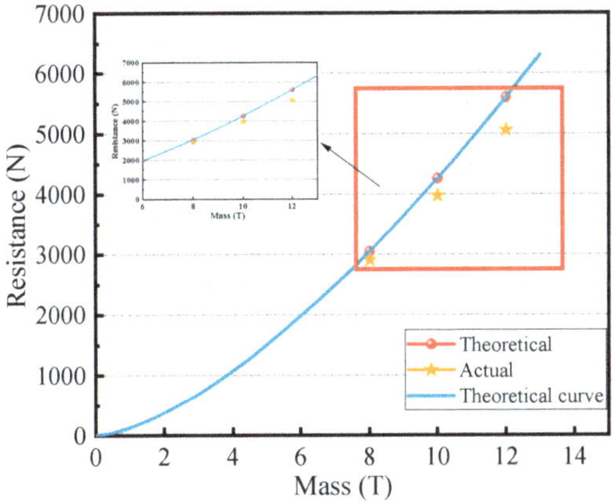

Figure 12. Comparison between theoretical values and actual values.

The table above shows that the deviation between the actual value and the theoretical value is less than 10%. The driving resistance affects the length of the soil runway. According to the calculation equation in reference [22], the error of the resistance is 10%, and the error in the sliding distance caused by the error is within the range of 50 m, which meets the requirements of the calculation accuracy of the runway length. This shows that the estimation method established in this paper meets the requirements of estimating the driving resistance of the soil pavement.

3.3.4. The Process of Estimating the Driving Resistance of Soil Pavement

Based on the idea of this paper, for a given soil runway, the process shown in Figure 13 can be used to estimate the resistance of a single wheel during aircraft operation.

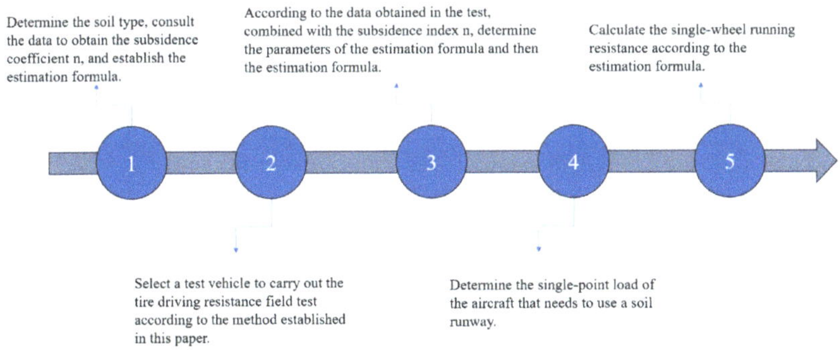

Figure 13. The estimation process of aircraft single-wheel running resistance on soil road surfaces.

4. Conclusions

Aiming at the complexity of determining wheel running resistance when an aircraft operates on a soil road surface, this study conducts a field resistance test based on tire operation characteristics and establishes an estimation method for aircraft tire running resistance. The following conclusions are drawn:

(1) Different methods can be used to characterize the driving resistance of the wheel on a soil road surface. Corresponding to different characterization methods, the driving resistance can be tested in two ways as follows: the stress state and the motion state.

(2) The tire resistance test method based on the motion state can better test the driving resistance of the tire on the soil road. This method can also be used to evaluate the load that the vehicle can bear on temporary soil roads.
(3) The resistance of the aircraft simulated load-loading vehicle is 2910 N, 3978 N, and 5055 N, respectively, when the single wheel load of 8T, 10T, and 12T is run on the soil pavement. The results obtained by the estimation method established in this paper are 3048 N, 4260 N, and 5600 N, and the deviations are 4.53%, 6.62%, and 9.73%, respectively. These small deviations indicate that our proposed resistance estimation method is feasible for practical application in estimating aircraft tire running resistance.

Author Contributions: Conceptualization, Z.W. and X.C.; methodology, C.L.; software, X.C.; validation, X.C.; formal analysis, Z.W.; investigation, G.W.; resources, X.C.; data curation, C.L.; writing—original draft preparation, Z.W.; writing—review and editing, X.C.; visualization, J.Z.; supervision, G.W.; project administration, J.Z.; funding acquisition, G.W. All authors have read and agreed to the published version of the manuscript.

Funding: This research received no external funding.

Institutional Review Board Statement: Not applicable.

Informed Consent Statement: Not applicable.

Data Availability Statement: Data is contained within the article.

Conflicts of Interest: The authors declare no conflict of interest.

References

1. Mataei, B.; Zakeri, H.; Zahedi, M.; Nejad, F.M. Pavement Friction and Skid Resistance Measurement Methods: A Literatu Review. *Open J. Civ. Eng.* **2016**, *6*, 537–565. [CrossRef]
2. Tan, Y.Q.; Xiao, S.Q.; Xiong, X.T. Testing Technology of Dynamic Road Friction. *J. Traffic Transp. Eng.* **2021**, *21*, 32–47. [CrossR
3. Gui, Z.; Liu, H.; Zhang, Z. Research Progress of pavement texture Characterization and anti-skid performance detection technol *Highw. Traffic Sci. Technol. Appl. Technol. Ed.* **2012**, *4*, 62–65.
4. Eubanks, J.J.; Haight, W.R.R.; Malmsbury, R.N.; Casteel, D.A. *A Comparison of Devices Used to Measure Vehicle Braking Decelerati* SAE Technical Paper; SAE: Detroit, MI, USA, 1993. [CrossRef]
5. Kordani, A.A.; Rahmani, O.; Nasiri, A.S.A.; Boroomandrad, S.M. Effect of Adverse Weather Conditions on Vehicle Braki Distance of Highways. *Civ. Eng. J.* **2018**, *4*, 46–57. [CrossRef]
6. Lee, Y.P.K.; Fwa, T.F.; Choo, Y.S. Effect of Pavement Surface Texture on British Pendulum Test. *J. East. Asia Soc. Transp. Stud.* **20** *6*, 1247–1257.
7. Zhang, P.; Cui, L.L.; He, L.; Xia, Q.S. Testing Technology of Dynamic Road Friction Coefficient. *Appl. Mech. Mater.* **2013**, *3* 249–253. [CrossRef]
8. Rizenbergs, R.L.; Burchett, J.L.; Napier, C.T. *Skid-Test Trailer: Description, Evaluation and Adaptation*; Commonwealth of Kentuc Department of Highways: Frankfort, KY, USA, 1972.
9. de Solminihac, H.; Chamorro, A.; Echaveguren, T. Procedure to Process, Harmonize and Analyze Grip Tester Measurements. Proceedings of the 7th International Conference on Managing Pavement Assets, TRB Committee AFD10, Calgary, AB, Cana 23–28 June 2008.
10. Rufford, P.G.; Gaughan, R.L. A Study of the Effect of Speed on the Skid Resistance of Fine and Coarse Textured Surfacings Usi the Mu Meter. In Proceedings of the 7th Australian Road Research Board (ARRB) Conference, Adelaide, Australia, 21–25 M 1974.
11. Tingle, J.S.; Norwood, G.J.; Cotter, B. Use of Continuous Friction Measurement Equipment to Predict Runway Condition Rati on Unpaved Runways. *Transp. Res. Rec.* **2017**, *2626*, 58–65. [CrossRef]
12. Ward, A.B.; Falls, A.J.; Rutland, C.A. Development of Deceleration-Based Friction Prediction Models and Methods Semiprepared Runway Surfaces. *Transp. Res. Rec. J. Transp. Res. Board* **2023**, *1*, 1–16. [CrossRef]
13. Ward, A.B.; Tingle Jeb, S.; Rutland, C.A. *Development of Deceleration-Based Runway Friction Measurement Methods*; US Arr Engineer Research and Development Center, Geotechnical and Structures Laboratory: Tampa, FL, USA, 2019.
14. Thompson, R.J.; Visser, A.T. Mine Haul Road Maintenance Management Systems. *J. South Afr. Inst. Min. Metall.* **2003**, *1* 303–312.
15. McAllister, M. Reduction in the Rolling Resistance of Tyres for Trailed Agricultural Machinery. *J. Agric. Eng. Res.* **1983**, 127–137. [CrossRef]

Rebati, J.; Loghavi, M. Investigation and Evaluation of Rolling Resistance Prediction Models for Pneumatic Tires of Agricultural Vehicles (Research Note). *Iran Agric. Res.* **2006**, *24*, 77–88.
Elwaleed, A.K.; Yahya, A.; Zohadie, M.; Ahmad, D.; Kheiralla, A.F. Effect of Inflation Pressure on Motion Resistance Ratio of a High-Lug Agricultural Tyre. *J. Terramechanics* **2006**, *43*, 69–84. [CrossRef]
Botta, G.F.; Tolon-Becerra, A.; Tourn, M.; Lastra-Bravo, X.; Rivero, D. Agricultural Traffic: Motion Resistance and Soil Compaction in Relation to Tractor Design and Different Soil Conditions. *Soil Tillage Res.* **2012**, *120*, 92–98. [CrossRef]
Bekker, M.G. Land Locomotion on the Surface of Planets. *ARS J.* **1962**, *32*, 1651–1659. [CrossRef]
Wong, J.Y. *Theory of Ground Vehicles*; John Wiley & Sons: Hoboken, NJ, USA, 2022; ISBN 1119719704.
Zhang, J.; Xu, W.; Quan, Z.Q.; Wang, J.; Li, J.Y.; Zuo, S.H.; Chen, T.Y.; Wang, Y.L. Design and application of aircraft load simulation loading vehicle. *Exp. Technol. Manag.* **2023**, *40*, 141–148+162. [CrossRef]
Wang, Z.H.; Cai, L.C.; Shao, B. Design Method of Military Airfield Runway Length. *J. Transp. Eng. Inf.* **2007**, *5*, 67–70+76.

coatings

Article

Assessment of *Warionia saharea* Essential Oil as a Green Corrosion Inhibitor for Mild Steel in HCl: Experimental and Computational Studies

Abdeslam Ansari [1], Youssef Youssefi [1], Mohamed Tanghourte [1], Nazih Ouassou [1], Nazih Asoufar [1], Mohamed Znini [1], Hassane Lgaz [2,*], El Houssine Mabrouk [1], Mohamed Azrour [1], Han-Seung Lee [3,*] and Belkheir Hammouti [4]

1 Laboratory of Materials Engineering for the Environment & Natural Resources, Faculty of Science and Technology, University Moulay Ismail of Meknes, BP 509, Errachidia 52003, Morocco; ansdesa2@yahoo.fr (A.A.); y.youssefi@edu.umi.ac.ma (Y.Y.); m.tanghourte@edu.umi.ac.ma (M.T.); n.ouassou@edu.umi.ac.ma (N.O.); n.asoufar@edu.umi.ac.ma (N.A.); m.znini@umi.ac.ma (M.Z.); mabrouk.elhoussine@gmail.com (E.H.M.); m.azrour@fste.umi.ac.ma (M.A.)
2 Innovative Durable Building and Infrastructure Research Center, Center for Creative Convergence Education, Hanyang University ERICA, 55 Hanyangdaehak-ro, Sangrok-gu, Ansan-si 15588, Gyeonggi-do, Republic of Korea
3 Department of Architectural Engineering, Hanyang University ERICA, 55 Hanyangdaehak-ro, Sangrok-gu, Ansan-si 15588, Gyeonggi-do, Republic of Korea
4 Euromed University of Fes, UEMF, Fes 30000, Morocco; hammoutib@gmail.com
* Correspondence: hlgaz@hanyang.ac.kr (H.L.); ercleehs@hanyang.ac.kr (H.-S.L.)

Abstract: The objective of this research work is the study of the inhibitory effect of *Warionia saharea* essential oil (WSEO) on the corrosion of mild steel (MS) in molar HCl solution, employing both experimental and theoretical methods. This inhibitory effect (IE) has been evaluated by using a combination of weight loss measurements (LW) and various electrochemical methods, such as open circuit potential (OCP), potentiodynamic polarization (PDP) and electrochemical impedance spectroscopy (EIS) experiments. The LW results indicated that IE increased with inhibitor concentration, reaching 83.34% at 3.00 g/L. The PDP analysis suggested that WSEO functions as a mixed inhibitor, while in the EIS results the R_{ct} values increased with inhibitor concentration to reach 165.8 Ω cm^2 at 2.00 g/L, suggesting a defensive film formation by WSEO molecules over the metallic surface. The thermodynamic study demonstrated that the WSEO molecules adsorption on the MS surface followed a Langmuir isotherm, involving mixed physical and chemical (physicochemical) adsorption on the MS surface. Theoretical methods, including density functional theory (DFT) and molecular dynamics (MD) simulations, were employed to elucidate the inhibition mechanisms of the three main components of WSEO. The quantum chemical analysis, using density functional theory (DFT) and molecular dynamics (MD) simulations, showed a low ΔE_{gap} value of 6.30 eV and a low adsorption energy (E_{ads}) value on an Fe (110) substrate of −258 Kcal/mol for (E)-Nerolidol, indicating the significant contribution of this molecule to the overall corrosion inhibition effect of WSEO. The scanning electron microscope (SEM) analysis verified the presence of a protective film formed by the inhibitor on the MS surface. This study highlights the potential of WSEO as a sustainable and green corrosion inhibitor in acidic environments.

Keywords: *Warionia saharea*; mild steel; corrosion inhibition; molecular dynamics

1. Introduction

The problems of deterioration of steel-based metal structures now occupy a fundamental place in the industry, particularly during descaling and acid pickling [1,2]. In this regard, effective means must be developed to control and/or prevent corrosion of these industrial parts. Currently, the exploration of components typically derived from natural sources, such as essential oils (EOs) extracted from various aromatic plants, as ecofriendly corrosion inhibitors, has proven to be an effective and promising way to reduce corrosion rates while minimizing economic losses and environmental and health impacts, because these alternatives are biodegradable, easy to extract and non-toxic [3,4]. This type of corrosion-inhibiting material has been proved effective in mitigating corrosion along with other inhibiting options such as carbon quantum dots [5]. For instance, the inhibitive effect of *Schinus mole* EO against the corrosion of carbon steel in 1.0 M HCl solution was reported and the corrosion inhibition performance was as high as 70% at 25 °C at 2 g/L [6]. The inhibitive action of Vietnam orange peel EO for MS in 1.0 M HCl solution has been investigated, and the maximum effectiveness (about 90%) was obtained at 3–4 g/L [6]. The EO of *Aaronsohnia pubescens* subsp. *pubescens* was reported as a green corrosion inhibitor for MS in 1.0 M HCl with a maximum performance of 89.88% at 3 g/L [7]. The inhibitory performance of these EOs results from their wide range of active centers such as functional groups (C=O, CO-O, O-H) and fragments composed of π electrons from double bonds and the aromatic cycle, which is capable of adsorbing on metal surfaces [3–8].

Environmentally friendly alternatives can have minimal health, environmental and safety risks and reasonably high protection efficiency at relatively low concentrations. The use of several environmentally friendly alternatives, including nanomaterials such as carbon dots (CDs) and their derivatives as well as plant extracts such as essential oils, has been extensively investigated as environmentally friendly alternatives to toxic corrosion inhibitors. However, in terms of preparation, essential oils are obtained via a simple and easy-to-use apparatus that only requires water as a solvent. For nanomaterials, their preparation requires nanomaterials, sometimes solvents, energy and time (i.e., multiple steps). In terms of application, essential oils have some disadvantages such as solubility in corrosive environments, supply (low yield) and mixing complexity. In addition, nanomaterials such as carbon dots have the following advantages: productivity, single-product synthesis, solubility (water as a solvent for preparation) and efficiency.

Furthermore, competitive interactions and synergistic effects between various molecules can coexist around the active regions of the metal surface, making it difficult to determine precisely which components present in the mixture can have a significant impact on corrosion suppression. However, it can be assumed that the main components are possible candidates that contribute to corrosion inhibition [7,9]. To confirm this hypothesis, it is desirable to study experimentally and individually the inhibitory power of each major component in corrosive environments, but this process is long and laborious, and sometimes the mechanism of inhibition remains unexplained. In this context, computational methods, such as density functional theory (DFT) and molecular dynamics (MD) simulations, have been developed as complementary approaches to explain complex "inhibitor/surface" interactions occurring in the corrosion process [10].

In an earlier investigation conducted in our laboratory, *Warionia saharea* essential oil (WSEO) was evaluated for its efficacy as an eco-friendly corrosion inhibitor for mild steel (MS) in a 0.5 M H_2SO_4 solution. The study revealed that WSEO achieved a maximum inhibition efficiency of 74% at a concentration of 3 g/L [11]. Building upon these findings, the present study focuses on assessing the corrosion inhibitory properties of WSEO on mild steel in a more aggressive 1 M HCl environment. We accomplished this by utilizing weight loss (WL) measurements in conjunction with electrochemical methods, specifically potentiodynamic polarization (PDP) and electrochemical impedance spectroscopy (EIS). These methods allowed for a comprehensive evaluation of WSEO's performance as a corrosion inhibitor under different conditions and provided deeper insights into its mechanism of action in acidic media.

The structure–activity relationship between the electronic/molecular properties of the main components and the WSEO efficiency was demonstrated by quantum chemical calculations using density functional theory (DFT). Molecular dynamics simulations were also carried out to describe the adsorption process of these molecules on the Fe (110) surface.

In parallel, a characterization of the surface of the steel was carried out using a scanning electron microscope (SEM).

2. Materials and Methods

2.1. Materials Preparation

Corrosion tests were carried out on MS samples having the following chemical composition (wt.%): P (0.09%), S (0.05%), Si (0.01%), Al (0.05%), Mn (0.21%) and the remainder iron. The corrosive medium was 1 M HCl prepared from a commercial HCl (37%) solution in distilled water.

The inhibitor used in this study was the WSEO extracted and characterized as described previously in our research [11]. It was obtained by hydrodistillation for 3 h using a Clevenger type apparatus with an average yield of 1.2%. Thirty-nine compounds, representing 93.2% of the total oil, were identified by gas chromatography (GC) and gas chromatography/mass spectroscopy (GC/MS). Figure 1 shows the names and structures of the three major components of WSEO: β-Eudesmol (34.9%), (E)-Nerolidol (23.0%) and Linalool (15.2%).

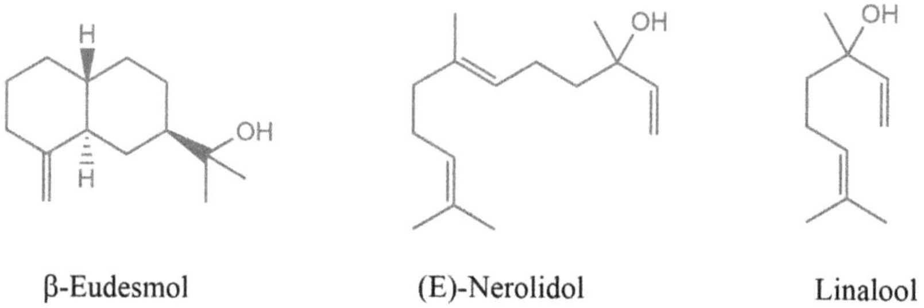

Figure 1. Chemical structures of the major components of WSEO.

2.2. Corrosion Tests

The inhibitory effects of *Warionia saharea* essential oil (WSEO) against MS corrosion in 1 M HCl were assessed using chemical and electrochemical methods, following the methodology previously established in our research [9]. To evaluate the inhibitor's protective efficacy, various concentrations of WSEO, ranging from 0.25 to 3.00 g/L, were tested. The essential oil was initially dissolved in pure ethanol to ensure uniform dispersion and effective interaction with the corrosive medium.

For the WL measurements, the inhibition efficiency (IE_{WL}) (%) was calculated based on Equation (1):

$$IE_{WL}\ \% = \frac{W_{corr} - W_{corr(inh)}}{W_{corr}} \times 100 \quad (1)$$

where W_{corr} and $W_{corr(inh)}$ are the values of the corrosion rate of MS in uninhibited and inhibited solutions, respectively.

For the electrochemical tests, firstly, the working electrode's open circuit potential (OCP) was tested for a period of 1800 s to achieve the stable corrosion potential (E_{corr}) value.

The inhibition efficiency (IE_{PDP} %) in PDP assay was calculated by the following Equation (2):

$$IE_{PDP}\ (\%) = \left(\frac{i_{corr} - i_{corr,\ inh}}{i_{corr}}\right) \times 100 \quad (2)$$

where i_{corr} and $i_{corr,\ inh}$ represent the corrosion current densities without and in the presence of the inhibitor, respectively.

In the case of the EIS test, the inhibitory efficiency (IE_{EIS}) (%) can be calculated using the following Equation (3):

$$IE_{EIS}\,(\%) = \left(\frac{R'_{ct} - R_{ct}}{R'_{ct}}\right) \times 100 \quad (3)$$

where R_{ct} and R'_{ct} are the charge-transfer resistance values without and with the inhibitor, respectively.

2.3. Scanning Electron Microscope (SEM) Analysis

SEM is generally employed to analyze microstructural and morphological changes in metal surfaces caused by corrosion. Specimens for characterization experiments were MS plates (3 mm × 3 mm × 1 mm). Three MS coupons were prepared in the laboratory by being rubbed with numerous abrasive papers ranging from 180 to 2000, then soaked in acetone as a degreasing solution, washed with double-distilled water, dried between two filter papers, resulting in a mirror finish, and stored in a desiccator.

Afterwards, two coupons were immersed in acidic solutions, both in the absence and presence of 2.00 g/L of the investigated inhibitor, at a constant temperature of 298 K for a duration of 24 h. The surface morphology of the metal was analyzed using scanning electron microscopy (JSM-IT500HR instrument (Tokyo, Japan)), operating at an accelerating voltage of 30 kV and with a magnification of about 20 µm.

2.4. Computational Studies

2.4.1. DFT Investigations

To correlate the obtained inhibition efficiency of WSEO with the molecular structure of its main compounds, some chemical reactivity indicators were calculated. Full geometry optimization was carried out using DFT calculations with the B3LYP level of theory and 6-311++ G (d,p) basis set [12]. Global quantum descriptors, such as E_{HOMO}, E_{LUMO}, and $\Delta E = E_{LUMO} - E_{HOMO}$ (Equation (4)), electronegativity (χ Equation (5)), overall hardness (η Equation (6)) and electron transfer proportion (ΔN Equation (7)), were calculated [13,14].

$$\Delta E_{gap} = E_{LUMO} - E_{HOMO} \quad (4)$$

$$\chi = \frac{E_{LUMO} - E_{HOMO}}{2} \quad (5)$$

$$\eta = \frac{E_{HOMO} - E_{LUMO}}{2} \quad (6)$$

$$\Delta N = \frac{\varnothing - \chi_{inh}}{2(\eta_{Fe} + \eta_{inh})} \quad (7)$$

In this context, $\varnothing$, which has a value of 4.82, represents the magnitude of the work function. The hardness of iron (Fe) is denoted as η_{Fe} = 0 eV/mol. The terms χ_{inh} and η_{inh} correspond to the electronegativity and hardness of the inhibitor, respectively. These parameters are essential for understanding the electronic interactions between the inhibitor molecules and the metal surface [15].

2.4.2. Molecular Dynamics Simulation

MD simulations were conducted to describe the adsorption of the three main components on metal surfaces. The simulation was conducted using the Forcite module in the Materials Studio 2020 software. The interactions within the studied system were modeled using a simulation box with dimensions of 24.82 Å × 24.82 Å × 34 Å. The energy surface was minimized using molecular mechanics, then the preliminary surface of 10 Å increased, and its periodicity was reformed by constructing a supercell. At last, the vacuum slab with a 26 Å thickness was constructed on the iron Fe (110) surface. The study was carried out in the solution, which is a mixture of 1 inhibitor molecule, 300 molecules of

H_2O, 20 molecules of H_3O^+, and Cl^- [16,17]. After the system achieves equilibrium, (E_{ads}), the energy of adsorption of the primary chemicals was computed. This was done using Equation (8) [18]:

$$E_{ads} = (E_{solution} + E_{total}) - (E_{inhibitor+solution} + E_{surface+solution}) \quad (8)$$

The mean square displacement (MSD) method is used to analyze the interactions between inhibitor molecules and corrosive species and find whether they affect the mobility and movement of particles inside the inhibitor film. In such a way, an amorphous cell module embedded in the Materials Studio 2020 program was used to generate the inhibitor film which held 50 parts (β-Eudesmol, (E)-Nerolidol, Linalool) monomer, 1 part Cl^- and 1 part H_3O^+ with the supercell size (23.450 Å × 23.45 Å × 23.45 Å) [19,20]. After the initial optimization, the MD equilibrium process was carried out, for 300 ps (1 fs time step) in a canonical ensemble (NVT) at 298 K. A Berendsen thermostat managed the system's temperature. All simulations used the COMPASS II force field [21,22].

3. Results and Discussion

3.1. WL Measurements

Table 1 provides a summary of the values of the corrosion rate (W_{corr}), inhibition efficiency (IE_{WL}%) and surface coverage ($\theta = IE_{WL}/100$) obtained by WL measurements in 1 M HCl with and without different doses of WSEO after 24 h of immersion at 25 °C.

Table 1. WL measurements of MS after 6 h immersion in blank and inhibited solutions by WSEO at 25 °C.

C_{WSEO} (g/L)	0.00	0.25	0.50	1.00	2.00	3.00
W_{corr} (mg·cm^{-2}h^{-1})	0.381	0.098	0.090	0.076	0.065	0.063
IE_{WL} (%)	-	74.15	76.36	79.96	82.93	83.34
Surface coverage (θ)	-	0.7415	0.7636	0.7996	0.8293	0.8334

The obtained results revealed that WSEO inhibits the MS's corrosion in a dose-dependent manner. With the addition of the inhibitor, W_{corr} is delayed and IE_{WL} (%) is significantly increased. Thus, IE_{WL} (%) goes from a value of 74.15% at 0.25 g/L to optimal and maximum values of 82.93 and 83.34% at 2.00 and 3.00 g/L, respectively. This characteristic explains the role of chemical composition of WSEO that is mainly electron-rich in inhibiting the corrosion [7,9]. In a previous study, the maximum value of IE_{WL} (%) of WSEO for MS corrosion in 0.5 H_2SO_4 solution was 74% at 3 g/L [11]. As a comparison, it can be noticed that the IE_{WL} (%) values of WSEO are higher in 1 M HCl compared to those in 0.5 H_2SO_4 throughout the concentration range tested, suggesting that the nature of the anions could influence the adsorption of the tested essential oil in acid solutions. This is consistent with previous studies that have demonstrated that the adsorption capacity of chloride ions (Cl^-) on the surface is more powerful than that of sulfate ions (SO_4^{2-}) [23,24]. Finally, the nature of the anions can affect the adsorption of the tested inhibitor in acidic solutions.

3.2. OCP Measurements

Figure 2 shows the evolution of the OCP during immersion of an MS in a 1 M HCl solution without and with different concentrations of WSEO at 298 K.

The results illustrated in Figure 2 clearly show that the potential (OCP) of the MS immersed in the 1 M HCl acid solution without and with inhibitor shows an increase to reach a quasi-stationary state at about 1800 s of immersion time, which indicates that the steel electrode/solution interface has reached a stable condition. Furthermore, with the increase in the concentration of inhibitors, a shift of OCP towards more positive values is noted. These variations can be justified by the formation of a layer on the surface of the electrode blocking the reactions that occur on the MS, such as the oxidation of iron and the reduction of hydrogen ions from hydrochloric acid [25].

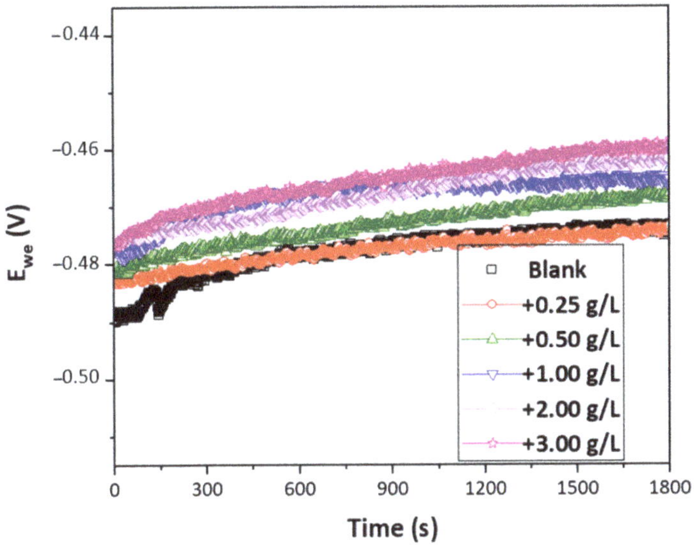

Figure 2. OCP–time curves for MS in 1 M HCl without and with various concentrations of WSEO.

3.3. *Potentiodynamic Polarization Curves*

PDP curves for MS in a 1 M HCl solution, both without and with various concentrations of WSEO, were measured at 25 °C, and the results are illustrated in Figure 3. The key polarization parameters, including corrosion current density (i_{corr}), corrosion potential (E_{corr}), cathodic Tafel slope (β_c), anodic Tafel slope (β_a) and inhibition efficiency (IE_{PDP}%), are detailed in Table 2. These parameters provide a comprehensive overview of the inhibitory performance and electrochemical behavior of WSEO in the acidic medium.

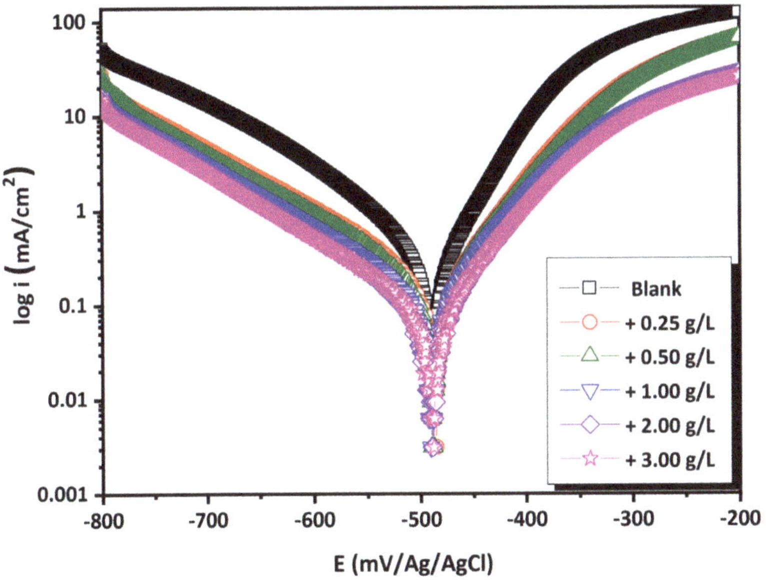

Figure 3. PDP curves of mild steel (MS) in 1.0 M HCl with and without varying concentrations of *Warionia saharea* essential oil (WSEO) at 25 °C.

Table 2. PDP parameters for mild steel (MS) in 1 M HCl with and without different concentrations of *Warionia saharea* essential oil (WSEO) at 25 °C.

C_{WSEO} (g/L)	$-E_{corr}$ (mV/SCE)	i_{corr} (μA cm^{-2})	$-\beta_c$ (mV dec^{-1})	β_a (mV dec^{-1})	IE_{PDP} (%)	θ
0.00	490	578.0	147.7	74.8	-	-
0.25	490	152.5	157.3	56.7	73.62	0.7362
0.50	490	130.6	149.8	53.6	77.40	0.7740
1.00	480	101.6	144.1	53.2	82.42	0.8242
2.00	480	92.4	140.8	49.9	84.01	0.8401
3.00	480	91.8	138.9	48.6	84.11	0.8411

Figure 3 demonstrates that the addition of WSEO significantly reduces both anodic and cathodic current densities. This reduction indicates a strong adsorption tendency of the oil on the MS surface, effectively delaying both anodic and cathodic reactions. These observations suggest that WSEO functions as a mixed-type inhibitor, providing comprehensive protection by impeding both the anodic and cathodic processes involved in corrosion [7,8]. The analysis of Table 2 clearly indicates that the IE_{PDP} (%) gradually increases with the addition of WSEO to the 1 M HCl solution until reaching its optimal value of 84.01% at 2.00 g/L. Further, the addition of the inhibitor changed the cathodic (β_c) and anodic (β_a) Tafel slopes in the inhibited system. This suggests that during the inhibition process, the inhibition mainly controls the cathodic and anodic reactions [7]. In addition, a slight shift of values of E_{corr} (below 85 mV) was detected, especially at high inhibitor concentrations, indicating the mixed character of the inhibitor and thus confirming the reduction of hydrogen evolution and metal oxidation [25].

3.4. EIS Measurements

Nyquist diagrams for MS corrosion in 1 M HCl solution and in the presence of different concentrations of the WSEO at 25 °C are shown in Figure 4a. The obtained EIS spectra were analyzed by fitting to the equivalent circuit illustrated in Figure 4b using EC-LAB V10.32 software, and the results are presented in Table 3.

Table 3. EIS values for MS in 1 M HCl with varying WSEO concentrations at 25 °C.

C_{WSEO} (g/L)	R_{ct} (Ω cm^2)	C_{dl} (μF cm^2)	CPE		IE_{EIS} (%)	θ	χ^2 (10^{-3})
			Q ($\Omega^{-1}S^n cm^{-2}$)	n			
0.00	21.71	73.37	314.51	0.877	-	-	2.25
0.25	69.1	52.52	219.23	0.865	68.58	0.6858	3.52
0.50	80.2	48.08	118.04	0.866	72.93	0.7293	5.40
1.00	134.8	42.78	96.70	0.864	83.90	0.8390	3.25
2.00	165.8	38.42	82.21	0.836	86.91	0.8691	5.25
3.00	166.1	37.64	82.11	0.834	86.92	0.8792	5.41

The Nyquist plots depicted in Figure 4a display a single capacitive loop for each test, suggesting that charge transfer is the primary mechanism controlling the corrosion processes in 1 M HCl [26]. This finding aligns with the single time constant observed in the Bode plots shown in Figure 4c,d. However, these capacitive loops are not ideal semicircles, which can be attributed to the dispersion effect commonly caused by the surface roughness and heterogeneity of the solid surfaces [27].

Moreover, as the concentration of the inhibitor increases, the width of the Nyquist plots also expands, indicating enhanced adherence of the natural substance to the mild steel (MS) surface. This increased width reflects the inhibitor's improved effectiveness in forming a protective barrier, thereby mitigating the corrosion process. The findings underscore the potent adsorption capabilities of the inhibitor, reinforcing its role in providing substantial protection against corrosion in acidic environments.

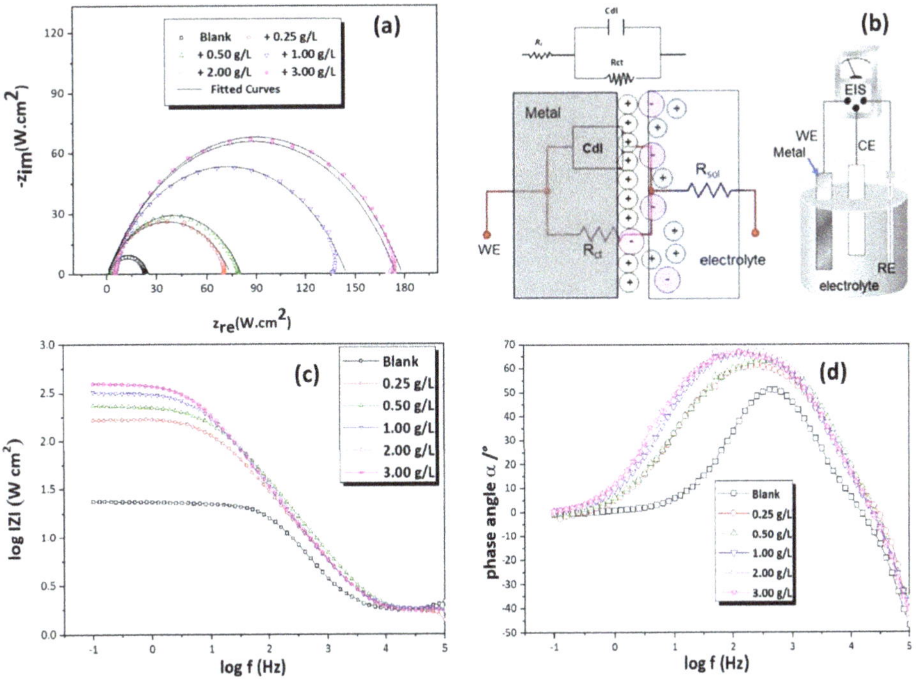

Figure 4. Nyquist diagrams (**a**), equivalent circuit (**b**) and Bode plots (**c**,**d**) of MS in 1 M HCl without and with the addition of various concentrations of WSEO at 25 °C.

Figure 4b shows the equivalent electrochemical circuit used to simulate the MS/acidic solution interface, where R_s is the solution resistance, R_{ct} is the charge transfer resistance and CPE is the constant-phase element to replace the double-layer capacitor (C_{dl}). The CPE element is employed to consider the heterogeneous surface of the working electrode and is defined by Equation (9) [2]:

$$Z_{CPE} = Q^{-1} \cdot (〚i \cdot w)〛^{-n} \quad (9)$$

where $i^2 = -1$, imaginary root; Q, CPE constant; ω, angular frequency (in rads^{-1}); n, deviation indicator which is a measure of surface inhomogeneity and its value lies between 0 and 1. n = 1 represents the ideal capacitor, n = 0.5 represents Warburg impedance, n = −1 represents inductance and n = 0 represents a resistor [27].

The fitting degree between the experimental data and the simulated results obtained by the proposed equivalent circuit was evaluated by the chi-square (χ^2) parameter (fit goodness).

The values of the double-layer capacitance (C_{dl}) were determined using the following formula of Equation (10) [3]:

$$C_{dl} = (Q \cdot R_{ct}^{1-n})^{1/n} \quad (10)$$

Based on Table 3, it can be seen that R_{ct} values rise with rising WSEO concentration. The highest R_{ct} value (166.1 Ω cm^2) has been obtained at 3.00 g/L, indicating the adsorption of the inhibitor and the development of a protective layer at the MS/solution interface that restricts the accessibility of Cl^- ions to the metal surface [28]. Alternatively, the C_{dl} value decreases with the addition of EO; it changes from 73.37 μF·cm^2 for the blank medium only to 37.64 μF·cm^2 for 3.00 g/L of WSEO. This diminution of C_{dl} values with the addition of the molecule can be explained by the rises in the dielectric constant and/or an increasing thickness of the electrical double layer.

Consequently, the IE_{EIS} values (%) show that the highest inhibitory performance is observed at the optimum concentration (2.00 g/L) and reach a maximum 87%. This result suggests that WSEO acts as a good adsorption inhibitor at the solution/metal interface forming a film or complex repelling corrosive species from the MS surface [14]. This behavior is in good agreement with that noticed from the WL and PDP results. Furthermore, the capacitance values of the double layer are also reduced to the maximum in the presence of the studied inhibitor compared to the uninhibited solution. The decrease in C_{dl} with increasing inhibitor concentration can be attributed to a decrease in the local dielectric constant and an increase in the thickness of the electrical double layer [28].

Moreover, the data in Table 3 clearly show that the values of the parameter χ^2 are very low on the order of 10^{-3}, which explains the agreement of the experimental impedance spectra with the proposed equivalent electrical circuit. In addition, it is noted that the surface heterogeneity parameter n changes with the addition of the inhibitor, indicating the surface modification when the WSEO is added [28]. In addition, the values of n remained relatively constant at around 0.8 (approaching unity), suggesting that the CPE is getting closer to the ideal capacitor behavior.

On the other hand, in a comparative study of the EO extracted from *W. saharea*, we note, according to Table 4, that the WSEO presents an excellent inhibitory performance in comparison with the other extracts of different and even low concentrations, which implies the importance of WSEO in the industrial sector.

Table 4. Comparison of WSEO inhibition efficiencies with those of other essential oils.

EO Inhibitor	Medium and Substrate	Concentration	IE (%)	Ref
Schinus mole	CS in 1 M HCl	2.00 g/L	70%	[6]
Vietnam orange peel	MS in 1 M HCl	4.00 g/L	about 90%	[7]
A. pubescens subsp. *pubescens*	MS in 1 M HCl	3.00 g/L	89.88%	[8]
Myrtle	Copper in 3% NaCl	10.00 g/L	91.88%	[27]
Rosemary			92.54%	
Asteriscus graveolens	MS in 0.5 M H_2SO_4	3.00 g/L	82.89%	[28]
Warionia saharea	MS in 0.5 M H_2SO_4	3.00 g/L	74%	[11]
Warionia saharea	MS in 1 M HCl	2.00 g/L	84%–87%	In this work

3.5. Adsorption Study

Adsorption isotherms are used to provide basic information about the nature of the interaction that takes place between the inhibitor and the metal surface. Indeed, to better describe the adsorption mode of WSEO molecules, attempts have been made to adjust the inhibitor concentration (C_{WSEO}) and the surface coverage values (θ) to certain adsorption isotherms such as the Langmuir isotherms, El-Awady, Freundlich and Temkin.

The linear equations (Equations (S1)–(S4)) of the studied isotherms are shown in Table S1 (Supplementary Materials) and their corresponding plots are represented in Figure 5. Depending on the model used, assumed differences can be taken into account such as the nature of the metallic surface and the interactions between the adsorbed molecules, as well as the structure of the layers formed by adsorbed species [29]. The correlation coefficient between surface coverage (θ) and the inhibitor concentration (C_{WSEO}) in the corroding solution were compared.

The equilibrium constant of the adsorption process K_{ads} is related to the Gibbs standard free energy of adsorption (ΔG°_{ads}) by the following relationship (Equation (11)) [6].

$$\Delta G^{\circ}_{ads} = -\,RT\ \ln\left(55.5\ K^{\circ}_{ads}\right) \quad (11)$$

where R is the universal gas constant, T is the thermodynamic temperature and 55.5 is the concentration of water in solution expressed in M (i.e., ~1000 g/L). Table 5 summarizes the estimated parameters from the studied isotherm graphs.

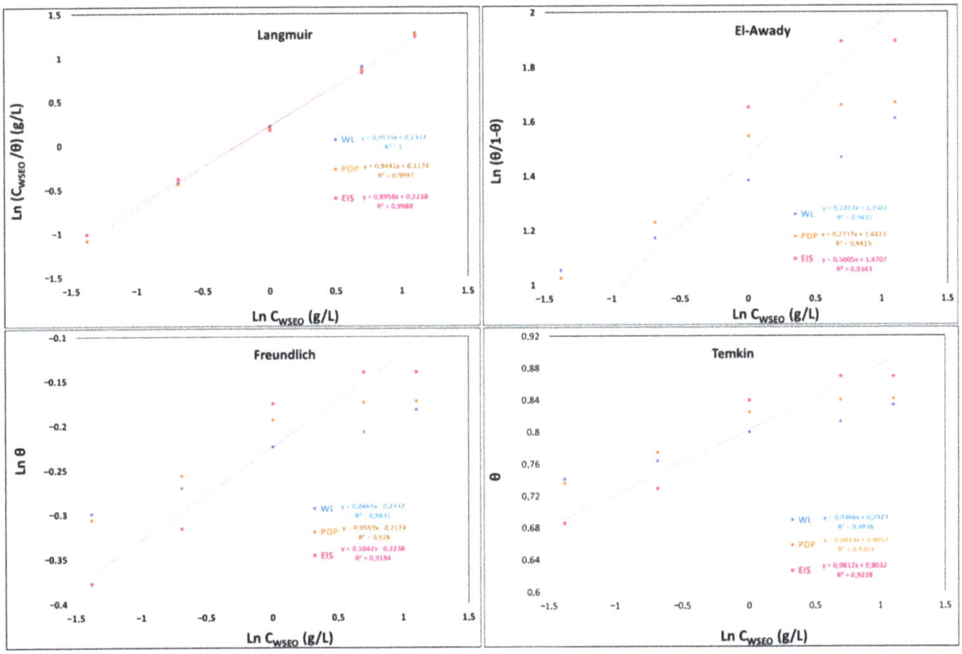

Figure 5. Isotherm models tested for MS in 1 M HCl in presence of WSEO using WL, PDP and EIS data at 25 °C.

Table 5. Thermodynamic finding for MS in 1 M HCl in the presence of WSEO at 25 °C.

Adsorption	Method	R^2	K_{ads} (L/g)	Isotherm Property	$\Delta G°_{ads}$ (KJ·mol^{-1})
Langmuir	WL	1	0.79	$R_L = 0.39$	−16.55
	PDP	0.9997	0.81	$R_L = 0.38$	−16.60
	EIS	0.9998	0.80	$R_L = 0.38$	−16.58
El-Awady	WL	0.9831	459.43	$1/y = 4.54$	−32.33
	PDP	0.9415	192.48	$1/y = 3.65$	−30.18
	EIS	0.9343	18.72	$1/y = 2.00$	−24.40
Freundlich	WL	0.9831	0.79	$z = 0.047$	−16.55
	PDP	0.926	0.80	$z = 0.056$	−16.58
	EIS	0.9194	0.80	$z = 0.104$	−16.58
Temkin	WL	0.9838	250×10^6	$a = -13.66$	−65.08
	PDP	0.9303	78.6×10^6	$a = -11.28$	−62.22
	EIS	0.9228	19.7×10^3	$A = -6.15$	−41.65

Examination of the plots presented in Figure 5 and the values listed in Table 5 revealed that the correlation coefficients (R^2) for all adsorption isotherms range from 0.91 to 1. It is clear that the corresponding plots of the Langmuir isotherm are straight lines with the correlation coefficients (R^2) close to unity ($R^2 > 0.999$) and that the values of the separation factor R_L are between 0 and 1, which confirms that the Langmuir isotherm model describes the appropriate adsorption of WSEO molecules on the MS surface in the acidic medium. However, the slope values are less than unity, suggesting that the real adsorption of the

WSEO did not strictly respect this isotherm. There may have been multilayer adsorption of inhibitor molecules or lateral interactions between the adsorbed species and the MS surface. According to the El-Awady adsorption parameter, the inverse of "*y*" obtained is between 2 and 4, suggesting that the WSEO molecule replaces up to four water molecules during the inhibition process schematized as follows (Equation (12)) [29,30].

$$Inh_{solution} + nH_2O_{electrode} \rightarrow Inh_{electrode} + H_2O_{solution} \quad (12)$$

Since the value of the adsorption parameter "z" in the Freundlich isotherm is less than 1, it is likely that the inhibitor studied will readily bind to the metal surface. Regarding the Temkin adsorbent isotherm, the values of the attractive parameter (a) are negative in all cases, showing that the lateral interaction is repulsive in the adsorption layer [31].

Generally, ΔG°_{ads} values lower than -20 KJ·mol^{-1} are linked to physical adsorption (physisorption), while ΔG°_{ads} values negatively greater than 40 KJ·mol^{-1} correspond to chemical adsorption (chemisorption). However, a physicochemical adsorption process (mixed adsorption) was observed when the ΔG°_{ads} displayed intermediate values between -20 KJ·mol^{-1} and -40 KJ·mol^{-1} [32]. In our study, ΔG°_{ads} of the inhibitor ranged between -16 and -65 KJ·mol^{-1}, indicating a physicochemical adsorption process onto the metal surface. Additionally, negative values of ΔG°_{ads} indicate spontaneous adsorption of molecules onto the MS surface, as well as strong interaction between active sites of molecules and the metal surface [4].

3.6. SEM Analysis

To gain deeper insights into the adsorption of WSEO molecules on the MS surface, we conducted a surface morphology analysis using scanning electron microscopy (SEM). Figure 6 presents the SEM images of mild steel (MS) immersed in 1 M HCl, both without and with 2.00 g/L of WSEO, after a 24 h exposure period.

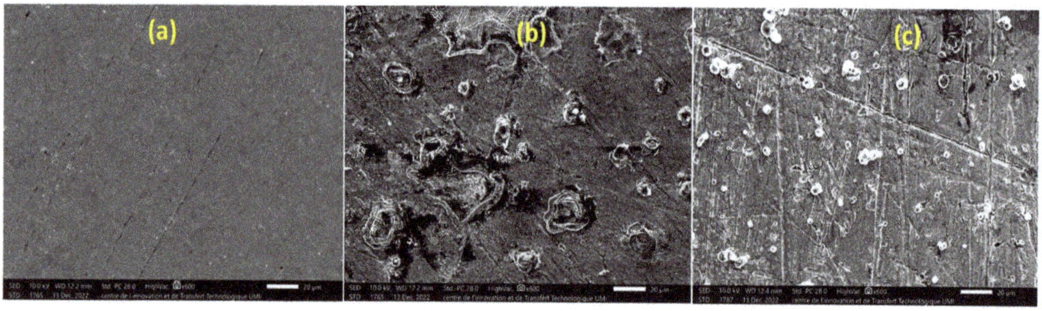

Figure 6. SEM morphology of MS surface (**a**) before immersion, (**b**) after 24 h immersion in blank solution and (**c**) after 24 h immersion in inhibited solution.

The morphology of MS after polishing and before exposure to the corrosive solution shows that the metal surface is clearly visible and has a characteristic associated with polishing scratches Figure 6a. After immersion in the corrosive solution without inhibitor, the SEM micrograph Figure 6b shows that the surface of the MS is seriously damaged due to the aggressive attack of the acid solution. However, by adding 2.00 g/L of WSEO, the surface morphology of the MS appears smoother and less corroded (Figure 6c). This behavior can be explained by the formation of a protective film adsorbed by WSEO molecules on the surface of the steel, which leads to a reduction in the corrosion process [33].

3.7. Theoretical Studies

3.7.1. DFT Calculations

Usually, EOs include various phytochemical components [5]. Because of this complexity, it is challenging to identify the components of an EO that could significantly contribute to preventing corrosion, but it can be assumed that phytochemicals with the highest percentages are responsible for EO performance [7,9]. In this regard, a DFT/B3LYP with 6-311++ G (d,p) basis set study was performed to correlate the inhibitive effect of the entire WSEO and the electronic properties of its major constituents (β-Eudesmol (34.9%), (E)-Nerolidol (23.0%) and Linalool (15.2%)).

Figure 7 illustrates optimized structures and LUMO-HOMO plots of β -Eudesmol, (E)-Nerolidol, and Linalool. Table 6 lists some global quantum chemical descriptors.

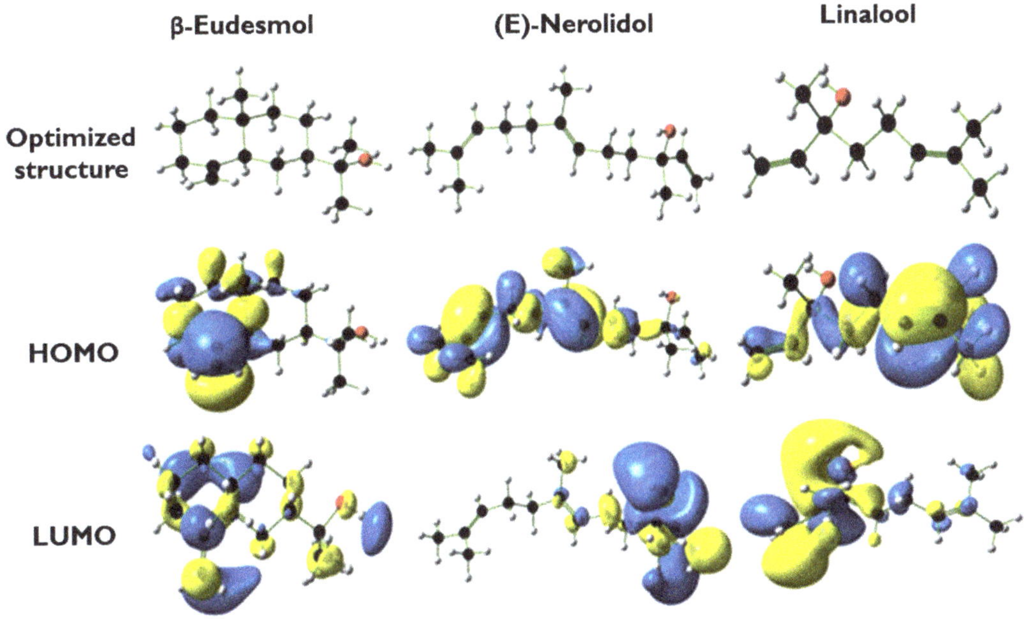

Figure 7. The HOMO-LUMO iso-surfaces and optimized geometry for β-Eudesmol, (E)-Nerolidol and Linalool that have been evaluated by the DFT/B3LYP with 6-311++ G (d,p) basis set.

Table 6. Calculated theoretical parameters for the major components of WSEO obtained from the DFT/B3LYP with 6-311++ G (d,p) basis set.

Molecule	E_{HOMO} (eV)	E_{LUMO} (eV)	ΔE_{gap} (eV)	ΔN	η
β-Eudesmol	−6.35	0.86	7.21	0.29	3.61
(E)-Nerolidol	−6.01	0.29	6.30	0.31	3.15
Linalool	−6.36	0.56	6.92	0.28	3.46

HOMO plots illustrate the regions of a molecule where electron donation is likely to occur, with a higher E_{HOMO} value signifying a greater propensity for the molecule to transfer electrons to the metal's unoccupied orbitals. Conversely, the density of the LUMO indicates the areas where the molecule can receive electrons, and a lower E_{LUMO} value denotes a stronger ability of the molecule to accept electrons from the metal surface [34]. Consequently, high E_{HOMO} values and low E_{LUMO} values favor the ability of the inhibitor to bind to the metal surface; thus, a low value of $\Delta E_{gap} = E_{LUMO} - E_{HOMO}$ increases the molecular reactivity and is expected to improve the inhibitory efficiency [35]. Figure 6 reveals that the HOMO and LUMO distributions for β-Eudesmol are similar, which were mostly situated around the methylcyclohexane motif. For (E)-Nerolidol, the HOMO density is mostly on the total molecule and is strongly spread around the (-C=C-) fraction, while the LUMO density is distributed over the atoms of the α, β-unsaturated hydroxyl group. The HOMO and LUMO orbitals for Linalool are similar and more intense along with the terminal double bond.

According to Table 6, (E)-Nerolidol has the highest E_{HOMO} and the lowest E_{LUMO} compared with β-Eudesmol and Linalool, which suggests that (E)-Nerolidol has the strong donor–acceptor properties. Likewise, the values of ΔE_{gap} and η follow the same theoretical order following (E)-Nerolidol < Linalool < β-Eudesmol, which shows that (E)-Nerolidol is more reactive than the β-Eudesmol. In addition, the ΔN value of (E)-Nerolidol is slightly higher than that of other molecules, which confirms the strong electron exchange of this molecule to and/or from the iron surface [36]. Based on the obtained findings, (E)-Nerolidol is expected to contribute to the corrosion inhibition of WSEO to a large extent, followed by Linalool and then β-Eudesmol.

3.7.2. MD Simulations

The optimized molecule–Fe (110) adsorption systems simulated by MD in the aqueous phase are shown in Figure 8. Correspondingly, the resulting adsorption energies calculated with MD simulation are listed in Table 7.

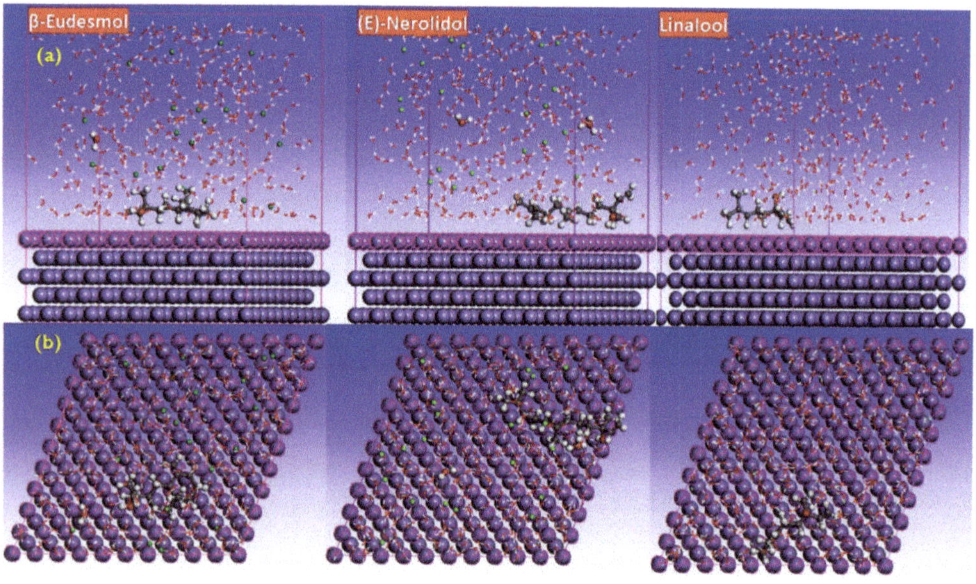

Figure 8. Top (**a**) and side (**b**) view of the most equilibrium adsorption configurations of the selected WSEO components on an Fe (110) surface obtained by MD simulations in the aqueous phase.

Table 7. The adsorption energies for the adsorption of the three main compounds of WSEO on the Fe (110) surface obtained by MD simulations (all in kcal/mol).

Molecule	E_{ads}
β-Eudesmol	−215.18
(E)-Nerolidol	−258.15
Linalool	−140.60

From Figure 8, it can be seen that the selected components of the WSEO are optimally adsorbed in a planar direction parallel to the surface of the Fe (110), thus maximizing the coverage of this surface thanks to the protective film created by the inhibitors causing the isolation of water molecules and other corrosive particles from the metal surface.

Moreover, the negative value of the adsorption energy indicates the exothermic and spontaneity of the adsorption process. Also, its high values show a stable and robust adsorption of the inhibitors on the Fe (110) surface, suggesting the rigidity of the interaction of inhibitors on the metal surface, thereby increasing their inhibition efficiency [37].

Table 7 indicates that the values of adsorption for selected molecules are negatives and obey the following order: (E)-Nerolidol > β-Eudesmol > Linalool.

According to Table 7, (E)-Nerolidol exhibits the highest adsorption energy, indicating its stronger affinity to adhere to the Fe surface forming a more stable protective layer against MS corrosion. Therefore, (E)-Nerolidol is the best candidate molecule to contribute substantially to the global corrosion inhibition effect of WSEO.

3.7.3. MSD Analysis

The MSD calculation was carried out in order to evaluate the anticorrosive performance of β-Eudesmol, (E)-Nerolidol and Linalool, by preventing the movement of H_3O^+ and Cl^- present in the corrosive solution. This model is based on the description of the migration rate of a corrosive species in inhibitor films by calculation of their diffusion coefficient (D_{ion}). Generally, the lower the value of the D_{ion} of corrosive species, the lower their migration, and thus the protective capacity of the inhibitor film will be higher [38]. Each supercell includes one cation of H_3O^+, one anion of Cl^-, and 30 molecules of the EO's main component. The diffusion coefficients of H_3O^+: D_{ion} (H_3O^+) and Cl^-: D_{ion} (Cl^-) resulting from the MSD curves can be determined by Einstein's equation (Equations (13) and (14)) [39,40]:

$$\mathrm{MSD}\,(t) = \frac{1}{N}\sum_{i=1}^{N} |R_i\,(t) - R_i(0)|^2 \tag{13}$$

$$D = \frac{1}{6}\lim_{t\to\infty}\frac{dMSD(t)}{dt} \tag{14}$$

where N reflects the number of diffusive atoms, while $R_i(0)$ and $R_i(t)$ translate the positions of corrosive ions at the origin time (0) and a later time (t), respectively.

A low value of D_{ion} means that corrosive species move poorly in the solution, thus preventing the electrochemical corrosion process. Conversely, a high value of D_{ion} suggests that the inhibitor film is weak [41]. Figure 9 displays MSD vs. time curves and the diffusion behaviors of D_{ion} (H_3O^+) and D_{ion} (Cl^-) in the supercell containing β-Eudesmol, (E)-Nerolidol and Linalool. In addition, the diffusion of corrosive species occurs via migration between free sites or cavities inside the inhibiting film on the metal surface. One method to examine the factors influencing the diffusion of corrosive species is through the calculation of fractional free volume (FFV). FFV is computed by Equation (15):

$$\mathrm{FFV}\,(\%) = \frac{V_f}{V_f + V_{oc}} \times 100 \tag{15}$$

where V_f is the free volume and V_{oc} is the volume occupied by the inhibitor film on the metal surface.

Typically, a high FFV value means that there are free cavities on the inhibition film in which corrosive species move easily from one cavity to another, resulting in a high diffusion coefficient and low efficiency of the inhibitor. Conversely, a low FFV value is associated with a low diffusion coefficient and high corrosion inhibition efficiency [41]. Figure 9 shows the molecular distribution used to calculate the V_f and V_{oc} volumes for systems built with a Connolly area. The calculated D_{ion} and the FFV values of both corrosive species in various inhibitor films are listed in Table 8.

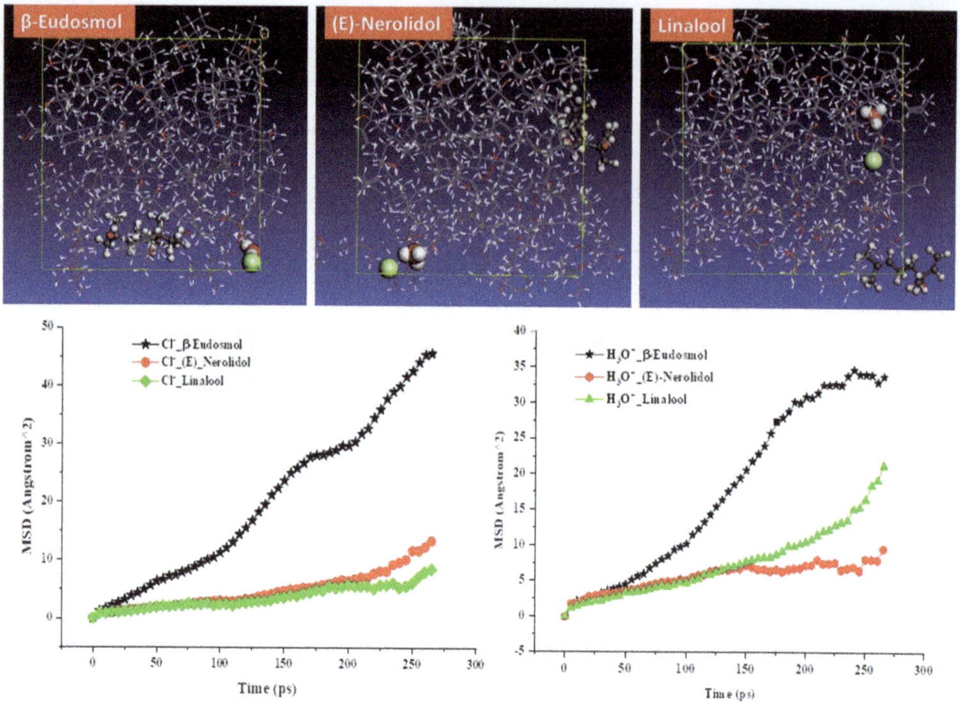

Figure 9. MSD vs. time curves and the diffusion behaviors of corrosive species (H_3O^+ and Cl^- ions) in three condensed inhibitor films at 25 °C.

Table 8. Calculated values of FFV and self-diffusion coefficient of the three molecules.

Molecule	D_{ion} (H_3O^+) ($m^2\ s^{-1}$)	D_{ion} (Cl^-) ($m^2\ s^{-1}$)	FFV (%)
(E)-Nerolidol	1.926×10^{-3}	1.062×10^{-2}	6.50
Linalool	1.57×10^{-2}	5.114×10^{-3}	6.94
β-Eudesmol	2.144×10^{-2}	3.153×10^{-2}	13.00

From Figure 9, the movement curve of corrosive species in β-Eudesmol is the most typical. It shifts by approximately 35 and 45 Å for H_3O^+ and Cl^-, respectively. However, in the case of (E)-Nerolidol and Linalool, the Cl^- shift curves are relatively smooth and weak (around 10 Å) while the H_3O^+ shift curves are characterized by a sharp jump in the case of Linalool (around 23 Å) at the end of the observation period. Therefore, based on the data in Table 8, we can classify the D_{ion} of H_3O^+ in all the inhibitor films examined as follows: D_{ion} (E)-Nerolidol) < D_{ion} (Linalool) < D_{ion} (β-Eudesmol). This order is in good accordance with the descriptors order reported by DFT calculations.

In Figure 10, we can see that the protective films formed by (E)-Nerolidol and Linalool are denser in the supercell; that is, the metal surface would be more covered than in the case of β-Eudesmol. In addition, from the FFV values (%), the sequence follows the same order as that of D_{ion}: (E)-Nerolidol < Linalool < -Eudesmol. This indicates that FFV decreases with molecular chain structure and length. It is well known that linear structures are more flexible than cyclic structures that are difficult to bend. Indeed, the increase in the linear chain strengthens the inhibition performance of (E)-Nerolidol ($C_{15}H_{26}O$) compared to Linalool ($C_{10}H_{18}O$). However, the bicyclic nature of ($C_{15}H_{26}O$) reduces its adsorption on the metal surface due to its rigidity. Obviously, the addition of these molecules to the solution limits the migration of corrosive species. Finally, the obtained findings confirm those discussed previously that (E)-Nerolidol may exhibit the highest reactivity among the three phytochemicals.

Figure 10. FFV of Cl^- and H_3O^+ in the four condensed inhibitor films at 25 °C.

3.8. Corrosion Inhibition Mechanism of WSEO

The first step in the suppression of the corrosion of metal is often the adsorption of inhibitors at the metal/solution interface. Such a mechanism might be sensitive to the chemical structure of the inhibitor and the metal's surface charge. As a result, the inhibitive activity of WSEO entails the adsorption of its active phytochemical components on the MS surface, particularly its main components namely β-Eudesmol (34.9%), (E)-Nerolidol (23.0%) and Linalool (15.2%). These compounds contain π-electrons of double bonds (C=C) and atomic oxygen in functional groups (O-H) that meet the normal adsorption centers. Moreover, extract organic components exist as neutral molecules and/or in the protonated form in the aqueous acid solution. Charged molecules (cations) can be absorbed by opposing charges via direct electrostatic interactions among the negatively charged molecules and the positively charged metal surface. Adsorbing may also be caused by no direct interaction between the protonated form and the chloride ions (Cl^-), primarily adsorbed on the MS surface charged positively (a physical mechanism). The neutral form can adsorb onto metal surfaces via a "donor-acceptor" interaction. This occurs between the non-bonded electron pairs of oxygen atoms and the π-electrons of double bonds, interacting with the "d" orbitals of the mild steel (MS) surface (a chemical mechanism) [4]. In this present study, E-Nerolidol was found to be the best candidate molecule that could significantly contribute to the overall corrosion inhibition effect of EOTL. Figure 11 illustrates the simplified scheme of the adsorption mechanism of E-Nerolidol on the C-steel surface in the 1.00 M HCl medium.

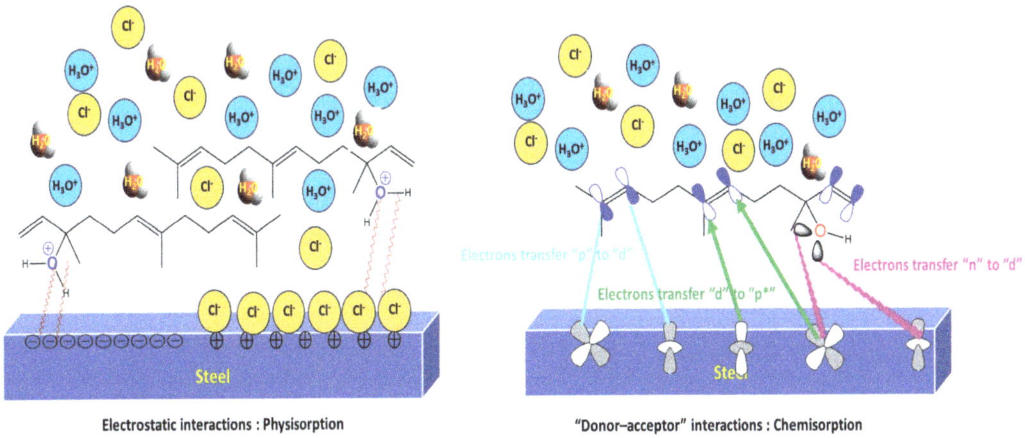

Figure 11. Simplified schema of the adsorption mechanism of (E)-Nerolidol molecules on the MS surface.

4. Conclusions

In this investigation, the inhibition ability of WSEO for MS corrosion in 1 M HCl medium was explored experimentally using weight loss (WL) measurement, electrochemical assays and SEM analysis, as well as theoretically using quantum DFT calculations and MC simulations. The key findings from these experiments are summarized as follows:

1. WSEO has been found to provide a satisfactory IE% of 86% at 2.00 g/L and 298 K, which suggests that it can be used as a good green corrosion inhibitor for MS in 1 M HCl medium.
2. The PDP study demonstrated that the examined inhibitor exerts a mixed inhibition effect and with increasing WSEO concentrations, the i_{corr} values dropped dramatically from 578.0 μA cm^{-2} in a blank solution to 92.4 μA cm^{-2} in the presence of 2.00 g/L, indicating significant corrosion inhibition potential.
3. Based on the EIS data, R_{ct} values rise with rising WSEO concentration reaching the optimal value of 165.8 Ω cm^2 at 2.00 g/L, indicating the adsorption of inhibitor and the development of a protective layer at the MS/solution interface.
4. The thermodynamic study revealed that the adsorption of WSEO on the MS surface aligns with the Langmuir isotherm model involving both chemical and physical adsorption.
5. DFT calculations provided a detailed explanation of the relationship between the inhibitory efficacy of WSEO and the electronic properties of its main constituents.
6. Monte Carlo (MC) simulations indicated that the primary compounds preferentially adsorb onto the Fe (110) surface in a flat, parallel orientation.
7. The MSD calculations reveal that the main component of the studied inhibitor creates a barrier film on the MS surface and limits the migration of corrosive species. SEM analysis confirms the existence of an inhibitor protective film on the MS steel surface.
8. The computational results also revealed that (E)-Nerolidol may play an important role in the overall inhibitory action of WSEO in 1 M HCl.

Supplementary Materials: The following supporting information can be downloaded at: https://www.mdpi.com/article/10.3390/coatings14091164/s1, Table S1. Linear equations and adsorption assumptions of the studied isotherms of MS in 1 M HCl for WSEO.

Author Contributions: A.A.: Conceptualization, Experiments, Formal Analysis, Resources, Data Curation and Writing the Manuscript; Y.Y., M.T., N.O. and N.A.: Conceptualization, Software, Investigation; E.H.M.: Methodology and Data Curation; M.A.: Validation, Software and Formal Analysis; M.Z.: Methodology, Visualization, Writing the Manuscript, Supervision and Review; H.-S.L. and H.L.: Conceptualization, Visualization, Investigation, Formal Analysis and Writing—Review and Editing; B.H.: Visualization, Supervision and Review. All authors have read and agreed to the published version of the manuscript.

Funding: This work was supported by the National Research Foundation of Korea (NRF), with a grant funded by the Korean government (MSIT) (No. NRF-2018R1A5A1025137).

Institutional Review Board Statement: Not applicable.

Informed Consent Statement: Not applicable.

Data Availability Statement: The data cannot be shared at this time as the data also form part of an ongoing study.

Conflicts of Interest: The authors declare no conflicts of interest.

References

1. Lavanya, D.; Priya, F.V.; Vijaya, D. Green Approach to Corrosion Inhibition of Mild Steel in Hydrochloric Acid by 1-[Morphol 4-yl (thiophen-2-yl) methyl] thiourea. *J. Fail. Anal. Prev.* **2020**, *20*, 494–502. [CrossRef]
2. Obot, I.; Meroufel, A.; Onyeachu, I.B.; Alenazi, A.; Sorour, A.A. Corrosion inhibitors for acid cleaning of desalination h exchangers: Progress, challenges and future perspectives. *J. Mol. Liq.* **2019**, *296*, 111760. [CrossRef]

Mzioud, K.; Habsaoui, A.; Ouakki, M.; Galai, M.; El Fartah, S.; Ebn Touhami, M. Inhibition of copper corrosion by the essential oil of *Allium sativum* in 0.5 M H_2SO_4 solutions. *SN Appl. Sci.* **2020**, *2*, 1611. [CrossRef]
Znini, M. Application of Essential Oils as green corrosion inhibitors for metals and alloys in different aggressive mediums. *Arab. J. Med. Aromat. Plants* **2019**, *5*, 1–34.
Sultana, N.; Hossain, S.; Mohammed, M.E.; Irfan, M.; Haq, B.; Faruque, M.; Razzak, S.; Hossain, M. Experimental study and parameters optimization of microalgae based heavy metals removal process using a hybrid response surface methodology-crow search algorithm. *Sci. Rep.* **2020**, *10*, 15068. [CrossRef]
Mellak, N.; Ghali, N.; Messaoudi, N.; Benhelima, A.; Ferhat, M.; Addou, A. Study of corrosion inhibition properties of *Schinus molle* essential oil on carbon steel in HCl. *Mater. Corros.* **2021**, *72*, 1270–1278. [CrossRef]
Manssouri, M.; Znini, M.; El Ouadi, Y.; Ansari, A.; Costa, J.; Majidi, L. Essential Oil of Aaronsohnia Pubescens Subsp. Pubescens as Novel Eco-Friendly Inhibitor for Mild Steel in 1.0 M HCl. *Anal. Bioanal. Electrochem.* **2020**, *12*, 841–856.
Bui, H.T.; Dang, T.-D.; Le, H.T.; Hoang, T.T. Comparative study on corrosion inhibition of vietnam orange peel essential oil with urotropine and insight of corrosion inhibition mechanism for mild steel in hydrochloric solution. *J. Electrochem. Sci. Technol.* **2019**, *10*, 69–81.
Znini, M.; Ansari, A.; Costa, J.; Senhaji, O.; Paolini, J.; Majidi, L. Experimental, quantum chemical and molecular dynamic simulations studies on the corrosion inhibition of c38 steel in 1M HCl by Anethum graveolens essential oil. *Anal. Bioanal. Electrochem.* **2019**, *11*, 1426–1451.
Saha, S.K.; Murmu, M.; Murmu, N.C.; Obot, I.; Banerjee, P. Molecular level insights for the corrosion inhibition effectiveness of three amine derivatives on the carbon steel surface in the adverse medium: A combined density functional theory and molecular dynamics simulation study. *Surf. Interfaces* **2018**, *10*, 65–73. [CrossRef]
Znini, M.; Majidi, L.; Laghchimi, A.; Paolini, J.; Hammouti, B.; Costa, J.; Bouyanzer, A.; Al-Deyab, S. Chemical composition and anticorrosive activity of Warionia saharea essential oil against the corrosion of mild steel in 0.5 M H_2SO_4. *Int. J. Electrochem. Sci.* **2011**, *6*, 5940–5955. [CrossRef]
Frisch, M. Gaussian. 2009. Available online: http://www.gaussian.com/ (accessed on 21 July 2023).
Umar, B.A.; Uzairu, A. In-silico approach to understand the inhibition of corrosion by some potent triazole derivatives of pyrimidine for steel. *SN Appl. Sci.* **2019**, *1*, 1413. [CrossRef]
El Ibrahimi, B.; Jmiai, A.; El Mouaden, K.; Baddouh, A.; El Issami, S.; Bazzi, L.; Hilali, M. Effect of solution's pH and molecular structure of three linear α-amino acids on the corrosion of tin in salt solution: A combined experimental and theoretical approach. *J. Mol. Struct.* **2019**, *1196*, 105–118. [CrossRef]
Murmu, M.; Saha, S.K.; Murmu, N.C.; Banerjee, P. Effect of stereochemical conformation into the corrosion inhibitive behaviour of double azomethine based Schiff bases on mild steel surface in 1 mol L^{-1} HCl medium: An experimental, density functional theory and molecular dynamics simulation study. *Corros. Sci.* **2019**, *146*, 134–151. [CrossRef]
Errahmany, N.; Rbaa, M.; Abousalem, A.S.; Tazouti, A.; Galai, M.; Touhami, M.E.; Lakhrissi, B.; Touir, R. Experimental, DFT calculations and MC simulations concept of novel quinazolinone derivatives as corrosion inhibitor for mild steel in 1.0 M HCl medium. *J. Mol. Liq.* **2020**, *312*, 113413. [CrossRef]
Haldhar, R.; Kim, S.-C.; Prasad, D.; Bedair, M.; Bahadur, I.; Kaya, S.; Dagdag, O.; Guo, L. Papaver somniferum as an efficient corrosion inhibitor for iron alloy in acidic condition: DFT, MC simulation, LCMS and electrochemical studies. *J. Mol. Struct.* **2021**, *1242*, 130822. [CrossRef]
Ouakki, M.; Galai, M.; Aribou, Z.; Benzekri, Z.; Dahmani, K.; Ech-chihbi, E.; Abousalem, A.S.; Boukhris, S.; Cherkaoui, M. Detailed experimental and computational explorations of pyran derivatives as corrosion inhibitors for mild steel in 1.0 M HCl: Electrochemical/surface studies, DFT modeling, and MC simulation. *J. Mol. Struct.* **2022**, *1261*, 132784. [CrossRef]
Berisha, A. Ab inito exploration of nanocars as potential corrosion inhibitors. *Comput. Theor. Chem.* **2021**, *1201*, 113258. [CrossRef]
Kalajahi, S.T.; Mofradnia, S.R.; Yazdian, F.; Rasekh, B.; Neshati, J.; Taghavi, L.; Pourmadadi, M.; Haghirosadat, B.F. Inhibition performances of graphene oxide/silver nanostructure for the microbial corrosion: Molecular dynamic simulation study. *Environ. Sci. Pollut. Res.* **2022**, *29*, 49884–49897. [CrossRef]
Benhiba, F.; Hsissou, R.; Abderrahim, K.; Serrar, H.; Rouifi, Z.; Boukhris, S.; Kaichouh, G.; Bellaouchou, A.; Guenbour, A.; Oudda, H. Development of New pyrimidine derivative inhibitor for mild steel corrosion in acid medium. *J. Bio-Tribo-Corros.* **2022**, *8*, 36. [CrossRef]
Benhiba, F.; Hsissou, R.; Benzekri, Z.; Echihi, S.; El-Blilak, J.; Boukhris, S.; Bellaouchou, A.; Guenbour, A.; Oudda, H.; Warad, I. DFT/electronic scale, MD simulation and evaluation of 6-methyl-2-(p-tolyl)-1,4-dihydroquinoxaline as a potential corrosion inhibition. *J. Mol. Liq.* **2021**, *335*, 116539. [CrossRef]
Deng, S.; Li, X. Inhibition by Ginkgo leaves extract of the corrosion of steel in HCl and H_2SO_4 solutions. *Corros. Sci.* **2012**, *55*, 407–415. [CrossRef]
Ngouné, B.; Pengou, M.; Nouteza, A.M.; Nanseu-Njiki, C.P.; Ngameni, E. Performances of alkaloid extract from *Rauvolfia macrophylla* Stapf toward corrosion inhibition of C38 steel in acidic media. *ACS Omega* **2019**, *4*, 9081–9091. [CrossRef] [PubMed]
Dhouibi, I.; Masmoudi, F.; Bouaziz, M.; Masmoud, M. A study of the anti-corrosive effects of essential oils of rosemary and myrtle for copper corrosion in chloride media. *Arab. J. Chem.* **2021**, *14*, 102961. [CrossRef]

26. Sathiya Priya, A.; Muralidharan, V.; Subramania, A. Development of novel acidizing inhibitors for carbon steel corrosion in 1! boiling hydrochloric acid. *Corrosion* **2008**, *64*, 541–552. [CrossRef]
27. Zarrouk, A.; Zarrok, H.; Ramli, Y.; Bouachrine, M.; Hammouti, B.; Sahibed-Dine, A.; Bentiss, F. Inhibitive properties, adsorpti and theoretical study of 3,7-dimethyl-1-(prop-2-yn-1-yl) quinoxalin-2 (1H)-one as efficient corrosion inhibitor for carbon steel hydrochloric acid solution. *J. Mol. Liq.* **2016**, *222*, 239–252. [CrossRef]
28. Pavithra, M.; Venkatesha, T.; Kumar, M.P.; Tondan, H. Inhibition of mild steel corrosion by Rabeprazole sulfide. *Corros. S* **2012**, *60*, 104–111. [CrossRef]
29. Bouklah, M.; Elmsellem, H.; Krim, O.; Serdaroğlu, G.; Hammouti, B.; Elidrissi, A.; Kaya, S.; Warad, I. Effect of substitution corrosion inhibition properties of three Imidazole derivatives on mild steel in 1 M HCl. *Arab. J. Chem. Environ. Res.* **2020**, 126–143.
30. Bockris, J.M.; Green, M.; Swinkels, D. Adsorption of naphthalene on solid metal electrodes. *J. Electrochem. Soc.* **1964**, *111*, 7 [CrossRef]
31. Wang, P.; Xiong, L.; He, Z.; Xu, X.; Hu, J.; Chen, Q.; Zhang, R.; Pu, J.; Guo, L. Synergistic Effect of Imidazoline Derivative a Benzimidazole as Corrosion Inhibitors for Q235 Steel: An Electrochemical, XPS, FT-IR and MD Study. *Arab. J. Sci. Eng.* **20** *47*, 7123–7134. [CrossRef]
32. Kowsari, E.; Payami, M.; Amini, R.; Ramezanzadeh, B.; Javanbakht, M. Task-specific ionic liquid as a new green inhibitor of mi steel corrosion. *Appl. Surf. Sci.* **2014**, *289*, 478–486. [CrossRef]
33. Zehra, B.F.; Said, A.; Eddine, H.M.; Hamid, E.; Najat, H.; Rachid, N.; Toumert, L.I. Crataegus oxyacantha leaves extract for carb steel protection against corrosion in 1M HCl: Characterization, electrochemical, theoretical research, and surface analysis. *J. M Struct.* **2022**, *1259*, 132737. [CrossRef]
34. Adewuyi, A.; Oluwaseyifunmi, A.; Kaki, S.S.; Oderinde, R.A. Synthesis of fatty phenylthiosemicarbazide from underutiliz *Sesamum indicum* seed oil: A promising corrosion inhibitor of carbon steel in developing country. *SN Appl. Sci.* **2019**, *1*, 6 [CrossRef]
35. Verma, C.; Obot, I.; Bahadur, I.; Sherif, E.-S.M.; Ebenso, E.E. Choline based ionic liquids as sustainable corrosion inhibit on mild steel surface in acidic medium: Gravimetric, electrochemical, surface morphology, DFT and Monte Carlo simulati studies. *Appl. Surf. Sci.* **2018**, *457*, 134–149. [CrossRef]
36. Salman, T.A.; Zinad, D.S.; Jaber, S.H.; Al-Ghezi, M.; Mahal, A.; Takriff, M.S.; Al-Amiery, A.A. Effect of 1,3,4-thiadiazole scaff on the corrosion inhibition of mild steel in acidic medium: An experimental and computational study. *J. Bio-Tribo-Corros.* **20** *5*, 48. [CrossRef]
37. Zhang, X.; Kang, Q.; Wang, Y. Theoretical study of N-thiazolyl-2-cyanoacetamide derivatives as corrosion inhibitor for aluminu in alkaline environments. *Comput. Theor. Chem.* **2018**, *1131*, 25–32. [CrossRef]
38. Hu, S.-Q.; Guo, A.-L.; Yan, Y.-G.; Jia, X.-L.; Geng, Y.-F.; Guo, W.-Y. Computer simulation of diffusion of corrosive particle corrosion inhibitor membrane. *Comput. Theor. Chem.* **2011**, *964*, 176–181. [CrossRef]
39. Singh, A.; Ansari, K.; Chauhan, D.S.; Quraishi, M.; Lgaz, H.; Chung, I.-M. Comprehensive investigation of steel corrosion inhi tion at macro/micro level by ecofriendly green corrosion inhibitor in 15% HCl medium. *J. Colloid Interface Sci.* **2020**, *560*, 225–2 [CrossRef]
40. Hsissou, R.; Benhiba, F.; Dagdag, O.; El Bouchti, M.; Nouneh, K.; Assouag, M.; Briche, S.; Zarrouk, A.; Elharfi, A. Developme and potential performance of prepolymer in corrosion inhibition for carbon steel in 1.0 M HCl: Outlooks from experimental a computational investigations. *J. Colloid Interface Sci.* **2020**, *574*, 43–60. [CrossRef]
41. Fernine, Y.; Arrousse, N.; Haldhar, R.; Raorane, C.J.; Kim, S.-C.; El Hajjaji, F.; Touhami, M.E.; Beniken, M.; Haboubi, K.; Tal M. Synthesis and characterization of phenolphthalein derivatives, detailed theoretical DFT computation/molecular simulati and prevention of AA2024-T3 corrosion in medium 3.5% NaCl. *J. Taiwan Inst. Chem. Eng.* **2022**, *140*, 104556. [CrossRef]

coatings

MDPI

icle

ear Thickening Polishing of Rotary Workpieces with Large urvature Edges

ndong Shang [1], Zewei Tang [2], Weiwei Zhang [1], Liqiang Wu [1], Jiahuan Wang [2,*] and Binghai Lyu [2,*]

1 Institute of Electronic Engineering, China Academy of Engineering Physics, Mianyang 621000, China
2 College of Mechanical Engineering, Zhejiang University of Technology, Hangzhou 310023, China
* Correspondence: wangjiahuan1994@foxmail.com (J.W.); icewater7812@126.com (B.L.)

Abstract: The polishing of rotary workpieces, which were characterized by large curvature edges, presents considerable challenges due to the difficulty of applying uniform polishing forces. Numerous solutions have been proposed to address this issue. However, none have successfully achieved an optimal balance between efficiency and surface quality. To this end, a "gentle" but effective method known as shear thickening polishing (STP) has been developed, which can improve the surface roughness to less than Ra 30 nm within a matter of minutes and without any surface abnormality. Influences of the STP were systematically investigated by single factor experimental designs, including abrasive type, abrasive concentration, installation angle, and polishing speed. And an illustrative case study involving the polishing of a rotary workpiece with a large curvature edge was presented, wherein the surface roughness can be reduced to Ra 25 nm within a mere 15 min and without the occurrence of burns or scratches typically associated with conventional polishing techniques. In contrast, traditional cashmere ball polishing requires approximately 30 min to achieve a similar surface quality, which often results in residual burns and scratches. The findings confirm that shear thickening polishing is an effective and efficient approach for enhancing the surface quality of rotary workpieces with significant curvature.

Keywords: large curvature edge; shear thickening polishing; surface roughness; 316L stainless steel

1. Introduction

Rotary workpieces with large curvature edges are characterized by their intricate surface shapes, which feature substantial curvature variations in various directions. These workpieces are essential in mechanical manufacturing and are crucial in sectors such as aerospace, national defense, military applications, biomedicine, and many others [1,2]. To meet the requirements of complex work environments, these surfaces often need polishing to improve their quality. A polished, high-quality surface with large curvature edges can greatly enhance the material's wear and corrosion resistance, thus prolonging its lifespan and boosting its overall performance.

Rotary workpieces with large curvature edges often have the characteristics of variable curvature themselves, and polishing tools need to make corresponding adjustments at different positions during the polishing process. In order to ensure the efficiency and uniformity of polishing, general polishing methods such as bonnet polishing [3], magnetorheological polishing [4], abrasive flow polishing [5], abrasive water jet polishing [6], and abrasive polishing wheel polishing have certain challenges. Bonnet polishing has the advantages of strong adaptability, low cost, and high processing efficiency [7], but it is limited by the curvature radius of the polishing head. For surfaces with varying curvatures, it is necessary to replace the polishing head multiple times to adapt to different curvatures. Since it is impossible to produce a very small-sized balloon processing tool, it is difficult to process large-curvature arc surfaces [8]. Magnetorheological polishing technology is widely used in surface processing due to its high processing stability, high adaptability to various curved surfaces, and low processing stress [9–11]. However, the high cost of designing the magnetorheological fluid and electromagnetic device limits its application space [12]. Abrasive flow polishing is suitable for processing complex surfaces, especially complex inner walls, due to its high efficiency and adaptability [6,13]. However, the accuracy and surface quality of the workpiece processed by abrasive flow cannot be effectively controlled, and problems such as "under-polishing" and "over-polishing" are common, which makes it difficult to meet the requirements of extremely strict shape accuracy and dimensional accuracy. This severely restricts the application of abrasive flow machining technology in a wider range [14]. Abrasive waterjet processing is highly flexible, simple to operate, and has no heat-affected zone, which makes it particularly suitable for processing deep-cavity structural parts and high-steepness complex curved surfaces. However, it has significant kinetic energy loss and unstable material removal functions [15,16], which make it difficult to ensure polishing efficiency and uniformity. It is mostly used for rough and semi-finishing processing. The abrasive polishing wheel polishing method allows for quick selection and replacement of abrasives based on different materials and polishing requirements, showcasing strong flexibility. Additionally, the polishing wheel can be reused, which reduces production costs. However, with the wear of the polishing wheel and uneven distribution of abrasives, it is prone to causing surface damage. Yang [17] utilized an abrasive cloth flap wheel to polish the surface of GH4169G and found that as the abrasive wheel wore down, deeper polishing tanks and burn marks formed on the surface of GH4169G, resulting in a decline in the quality of the polished surface.

According to the shear thickening effect of non-Newtonian fluid, when there is relative movement between the polishing slurry and the workpiece, the surface area of the polishing slurry and the workpiece will produce shear thickening phenomenon due to the action of shear stress, forming a "flexible fixed mold" on the surface of the curved surface so as to achieve material removal on the surface. The shear thickening polishing (STP) method has the ability to fit complex surfaces well and can ensure the polishing requirements of complex surfaces [18]. Shao [19] has successfully realized efficient and high-quality polishing of turbine blades of high temperature nickel-based alloys by using the principle of shear rheological properties. After processing with a linear polishing speed of 1.88 m/s for 12 min, blade surface roughness Ra rapidly decreased from 73 nm to 8.5 nm. After further process optimization, the final result is a high-quality surface of Ra 4.5 nm. Wang [20] carried out shear thickening polishing passivation for carbide milling cutters. After passivation and polishing, the cutting edge became rounded, the front and back surfaces became flat, and the defects of the cutting edge were removed. Nguyen [21] applied the characteristics of shear thickening fluid (STF) to the machining of alloy steel SCM435 gear and obtained the best surface with a roughness of 13 nm by using a multi-response optimization method. Wang [22] applied shear thickening polishing to PCB microdrills, and the cutting-edge surface of the micro-drills was improved, resulting in a significant increase in tool life. Zhou [23] combined the shear-thickening effect with electrorheological fluid for polishing alumina ceramic workpieces, resulting in a reduction of surface roughness from Ra76.3 nm to Ra16.0 nm. Zhu [24] proposed an effective shear thickening micro-polishing method with fast tool servo control for micro structured surfaces.

These surfaces achieved higher surface smoothness and conformed to the desired surface shapes after polishing. Zhou [25] employed shear thickening polishing technology to perform edge preparation on PCB microdrills, resulting in an increase in the radius of the main cutting edge from an initial 2.77 ± 0.4 μm to 23.9 ± 0.3 μm.

However, the shear-thickening polishing process for rotary workpieces with large curvature edges has not been systematically summarized and elaborated. In this paper, the shear thickening polishing (STP) method is used to carry out polishing tests on rotary workpieces with large curvature edges. The single factor experimental method is used to study the influence of abrasive particle type, abrasive particle concentration, installation angle, and polishing speed on the arc section of a rotary workpiece with a large curvature edge. The optimal polishing parameters are obtained through process optimization, and the polishing experiments are verified.

2. Principle of Shear Thickening Polishing (STP)

The shear thickening polishing (STP) method is a new non-contact ultra-precision polishing method, which utilizes the nonlinear rheological properties of non-Newtonian polishing slurry under shear stress to achieve material removal. The basic principle of shear-thickening polishing is shown in Figure 1. The fluid phase of shear thickening polishing with good fluidity forms a continuous flow field on the surface of the workpiece, and the interface shear stress generated by the contact interface between the workpiece and the flow field causes rheological effects such as local thickening and solidification of the polishing slurry to form a rheological layer that holds the abrasive particles in a flexible way. The rheological layer can achieve surface polishing by micro-cutting the micro convex peaks on the workpiece surface. As a new flexible polishing method, the shear thickening polishing method can realize the precision polishing of the surface of a rotary workpiece with a large curvature edge. Shear thickening slurry was adaptable to different complex structures with large curvature. When the polishing slurry flows over the large curvature edge, it continuously circulates to form a stable shear flow field, enabling efficient polishing of rotary workpieces with large curvature edges.

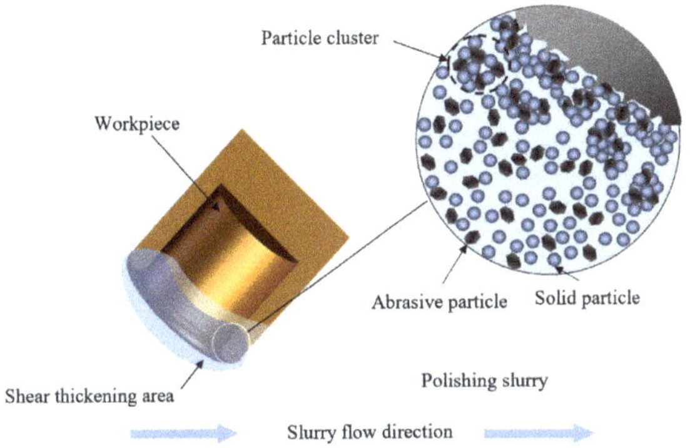

Figure 1. Schematic illustration of the STP.

3. Experimental Method and Conditions

3.1. Preparation of STP Slurry

Shear thickening fluid (STF) is one of the key factors that determine the efficiency of shear thickening polishing and the resulting surface quality. Shear thickening fluid (STF) is composed of dispersed phase particles of poly-hydroxyl polymer (PHHP), deionized water, and abrasive particles. Prepare deionized water at a certain temperature (30–50 °C), gradually add PHHP, and dissolve while stirring to form a shear-thickening fluid. After standing for a period of time, the abrasive particles are gradually added, stirred evenly, and dispersed by an ultrasonic device for 15 min to form STF.

The stress-controlled rheometer can measure the rheological curve of shear-thickened polishing slurry. The test temperature is set at 25 °C. The rotor has a diameter of 0.25 mm and a cone angle of 2°. Figure 2 shows the rheological characteristics of shear-thickened polishing slurry with silica sol as an abrasive particle at different pH values [22]. With the increase in shear rate, the viscosity will show a sudden upward trend within a certain range. The pH of the shear thickening fluid in this paper is not adjusted, and its pH is between 8 and 9. Figure 2 roughly reflects its rheological trend.

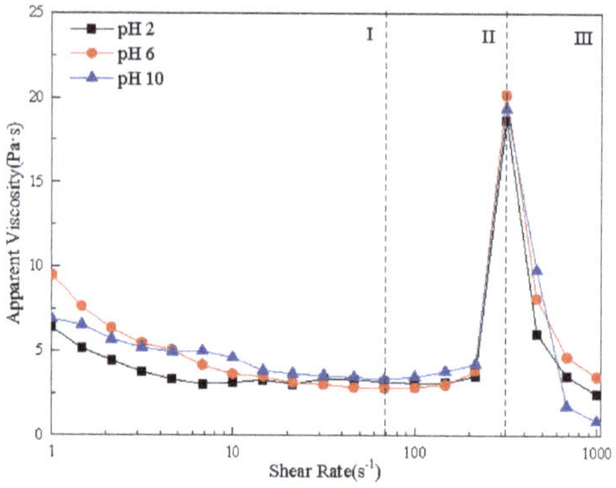

Figure 2. The viscosity curve of STF [26].

3.2. Experimental Method and Conditions

The experimental device is shown in Figure 3b. The rotary workpiece with a large curvature edge was immersed in slurry and rotated along the Z axis to ensure the uniform polishing of the surface. The distance of the workpiece from the edge of the polishing tank is 5 mm. The polishing tank drives the shear thickening polishing slurry to rotate, making the workpiece and the polishing slurry move relative to each other, resulting in the shear thickening effect to achieve polishing, as shown in Figure 3a. The single factor experiment method was used to study the influence of abrasive particle type, abrasive particle concentration, installation angle, and polishing speed on the circular arc section of the rotary workpiece with a large curvature edge. The test conditions are shown in Table 1.

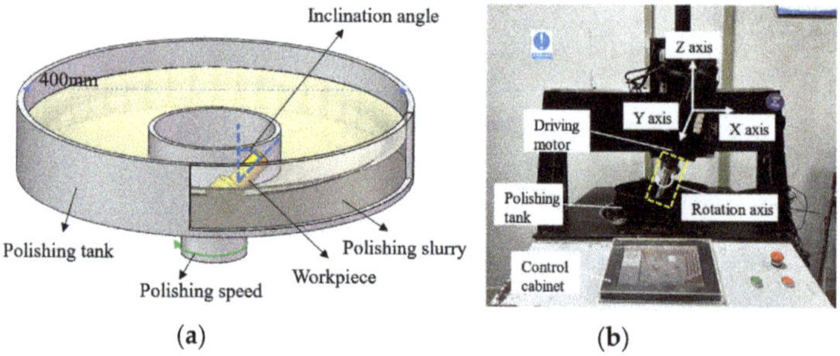

Figure 3. (**a**) Schematic illustration of STP. (**b**) Polishing device employed in this study.

Table 1. Experiment conditions.

Experimental Conditions	Parameters
Diameter of polishing tank (mm)	400
Rotation speed of samples (rpm)	10
Abrasive	Diamond, Alumina, Colloidal silica
Diamond concentration (wt.%)	3, 6, 9, 12
Abrasive size (nm)	80
Polishing speed (rpm)	45, 50, 55, 60
Installation angle (°)	45, 50, 55, 60
Time per trial (min)	15

In this study, rotary workpieces with large curvature edges were shear thickened and polished. The rotary workpiece with a large curvature edge was 316L stainless steel, and its main physical properties and basic parameters are shown in Tables 2 and 3. The main feature of this part is a complex arc surface at the top, which is composed of three small arcs with extremely large curvatures. Its specific morphology is shown in Figure 4.

Table 2. Properties of 316L stainless steel [27,28].

σ_b/MPa	$\sigma_{0.2}$/MPa	δ_5/%	ψ/%
$\geq$480	$\geq$177	$\geq$40	$\geq$60

Table 3. Parameters of workpiece.

Initial Conditions	Values
Material	316L stainless steel
Radius of the top arc r1 (mm)	2
Radius of the inner arc r2 (mm)	2.1
Radius of the outer arc r3 (mm)	2.2
Roughness of outer arc Ra/Sa (nm)	63.9 ± 5.7/36.3 ± 2.4
Roughness of top arc Ra/Sa (nm)	83.4 ± 4.3/40.9 ± 3.1
Roughness of inner arc Ra (nm)	78.0 ± 4.4

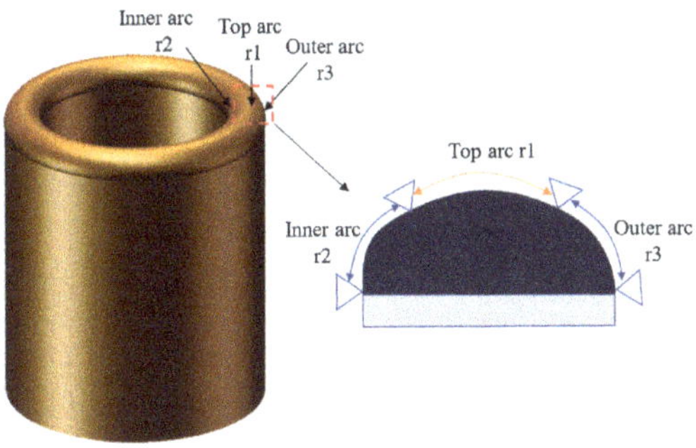

Figure 4. The specific topography of the workpiece.

3.3. Measurement Method

The 3D surface topography of the workpiece was observed with the optical 3D surface profiler, and the Sa of each surface roughness was observed with a 50× lens. The macroscopic surface morphology of the workpiece was observed using an ultra-depth field microscope. A profilometer with a resolution of 0.1 nm was used to measure the roughness

Ra of each position. We measured the roughness values of the arc segments with the Form Talysurf i-Series profilometer and analyzed the data with the software Ultra 6.1.12.1, which comes with the machine. The roughness filter was selected in G mode, and the length of the measured sampled data was 1.25 mm. Five points of the inner arc, top arc, and outer arc were measured, respectively, and the measurement diagram of each position was shown in Figure 5.

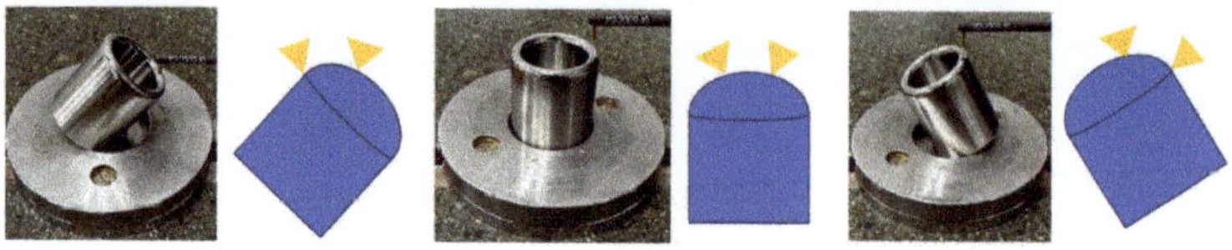

(**a**) (**b**) (**c**)

Figure 5. Schematic of workpiece measurement: (**a**) Top arc; (**b**) Inner arc; (**c**) Outer arc.

The arc of the rotary workpieces with a large curvature edge is a complex surface with a small size and large curvature, which is easy to fit distortion in the measurement process and needs to be corrected as shown in Figure 6.

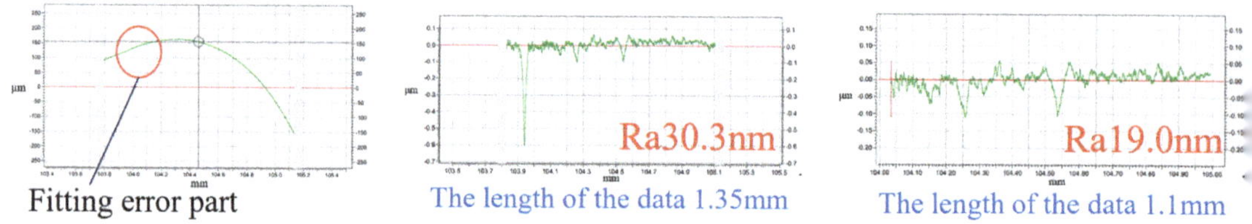

(**a**) (**b**) (**c**)

Figure 6. Schematic of fitting correction: (**a**) Fitting distortion; (**b**) Fit distorted results; (**c**) Fit correction results.

4. Experimental Results and Analysis

4.1. Effect of Abrasive Particle Type on Surface Quality

To avoid scratches on curved surfaces during polishing while also considering polishing efficiency, the selection of abrasive particles is prioritized. Before obtaining effective polishing parameters, the top arc, which is most easily accessible to the polishing slurry, is polished first. The suitability of the abrasive particle is assessed by observing the surface roughness and morphology of the top arc.

With the polishing rotational speed controlled at 55 r/min, the self-rotation speed at 10 r/min, the installation angle at 0°, and the polishing time set to 15 min, different abrasive particles were studied for polishing the top arc, with the results shown in Figures 7 and 8. As time increases, silica sol exhibits the best effect in improving surface roughness, while alumina and diamond perform similarly. Compared to alumina and diamond, silica sol exhibits a slower rate of improvement in surface quality, but it does not damage the workpiece surface. Therefore, as time increases, the surface quality continues to improve. However, for alumina and diamond, they can significantly reduce roughness when the surface quality of the workpiece is poor. However, as the surface quality improves, the hardness of these two abrasives leads to the formation of holes and scratches, causing the roughness to increase.

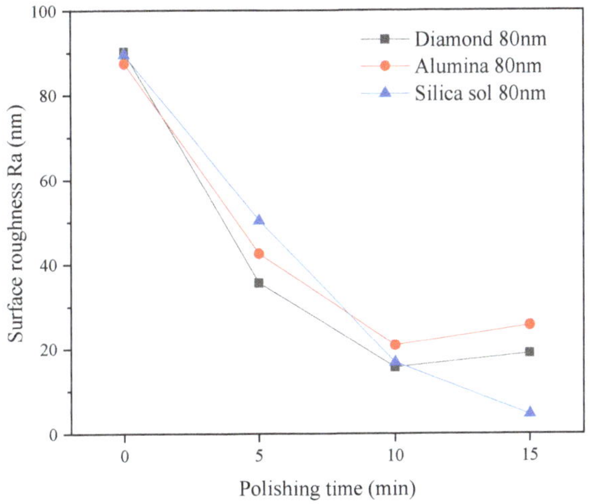

Figure 7. Effect of abrasive particle type on top arc roughness Ra.

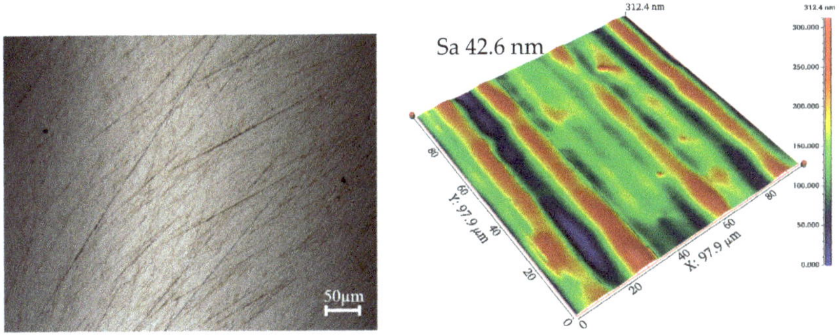

(**a**)

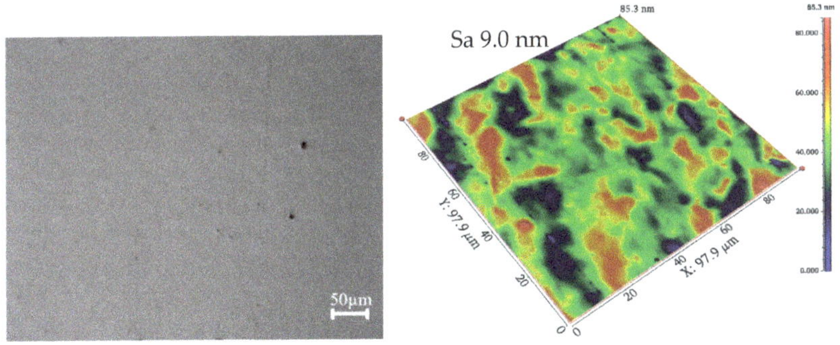

(**b**)

Figure 8. *Cont.*

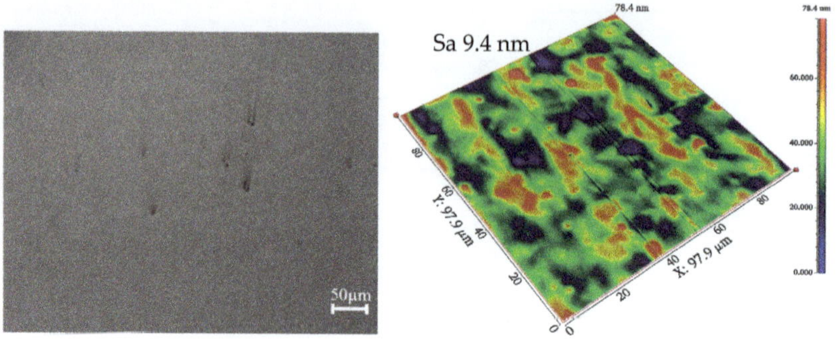

(**c**)

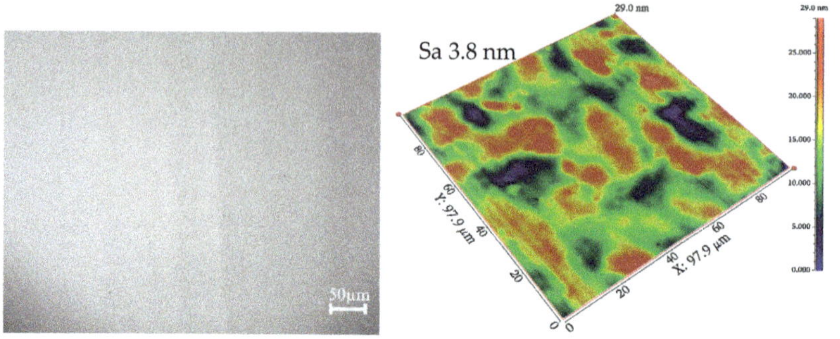

(**d**)

Figure 8. Surface topography of top arc polished by different abrasive particles: (**a**) Initial surface Sa 42.6 nm; (**b**) Diamond 80 nm Sa 9.0 nm; (**c**) Alumina 80 nm Sa 9.4 nm; (**d**) Silica sol 80 nm Sa 3.8 nm.

4.2. Effect of Abrasive Particle Concentration on Surface Quality

The influence of abrasive particle concentration on shear-thickened polishing slurry is mainly reflected in the difference in rheological properties. Generally, the higher the abrasive particle concentration, the lower the critical shear rate of polishing slurry and the higher the material removal rate.

Silica sol with a particle size of 80 nm was used as an abrasive particle. The polishing speed was controlled at 55 r/min, the rotation speed was 10 r/min, and the polishing time was 15 min. The roughness value was measured every 5 min. The relationship between the top arc surface roughness value and the abrasive particle concentration with the increase in time is shown in Figure 9. When the concentration of silica sol is 9 wt.%, the decreased rate of roughness and the final roughness value are good, and the roughness Ra can reach 24.1 nm. As the concentration further increases to 12 wt.%, the change in roughness is not significant.

The higher the abrasive concentration means that there will be more silica sol abrasive particles in the same polishing area, the more effective abrasive particles involved in polishing, the more the roughness peak of the workpiece surface will be removed, and the faster the surface roughness value will decline. After polishing the top arc position with a polishing slurry of different particle concentrations for 15 min, the surface morphology of the top arc is shown in Figure 10.

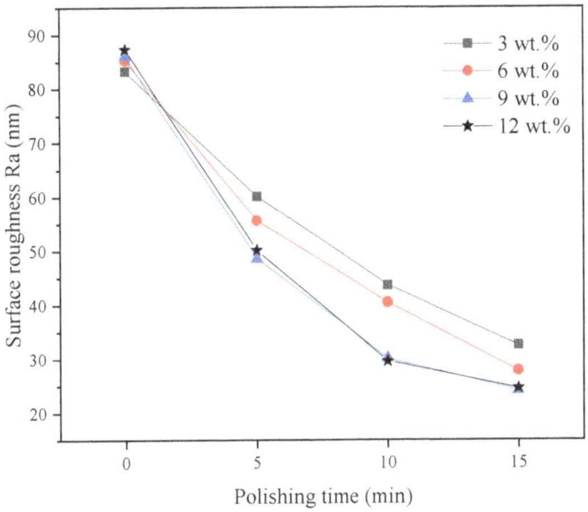

Figure 9. Effect of abrasive particle concentration on top arc roughness Ra.

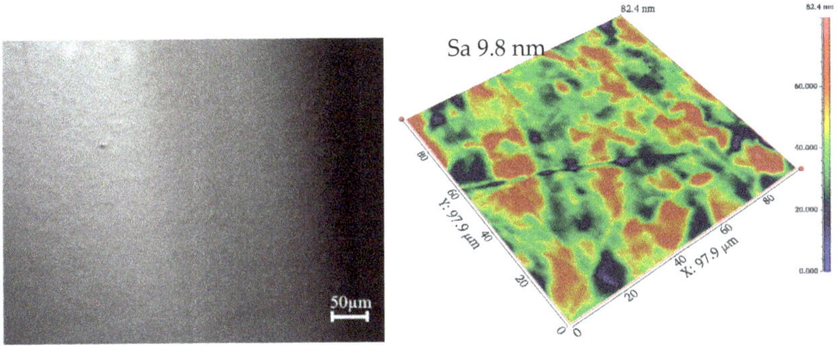

(**a**)

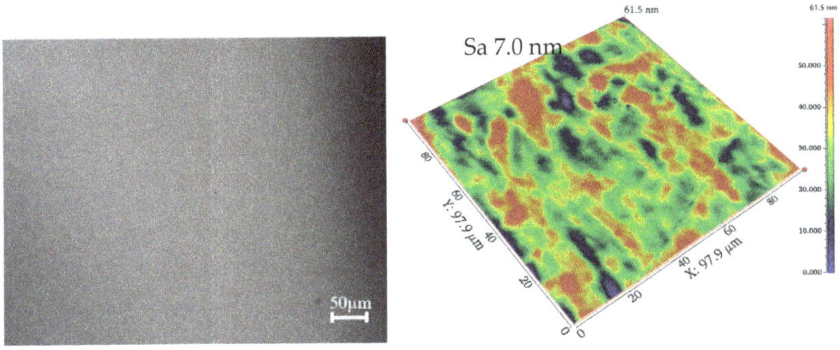

(**b**)

Figure 10. *Cont.*

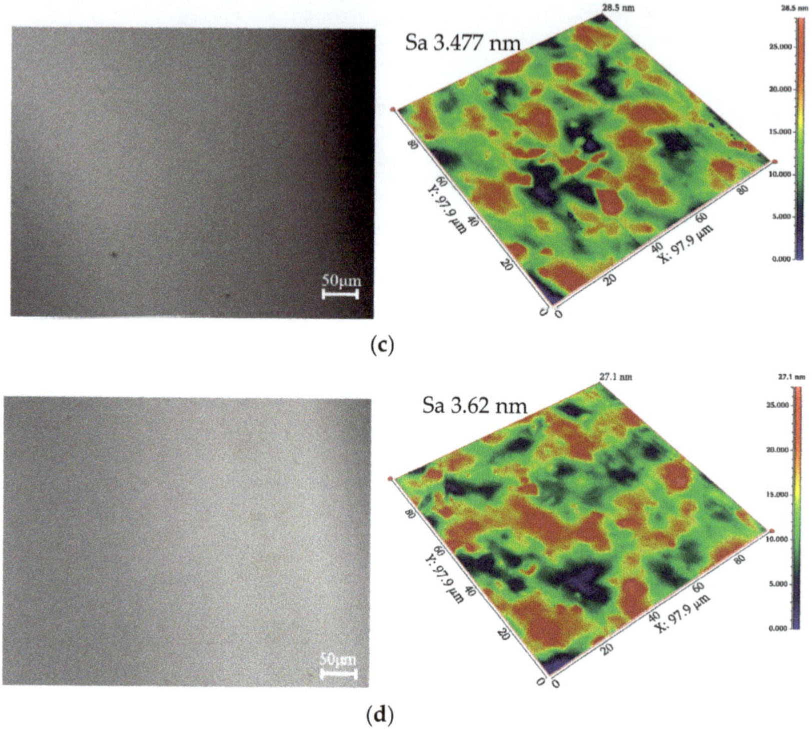

Figure 10. Surface morphology and roughness measurement results after polishing the top arc with different abrasive concentrations: (**a**) 3 wt.% Sa 9.8 nm; (**b**) 6 wt.% Sa 7.0 nm; (**c**) 9 wt.% Sa 3.477 nm; (**d**) 12 wt.% Sa 3.62 nm.

4.3. Effect of Installation Angle on Surface Quality

Considering that one part of the arc section is the inner arc, and due to the polishing principle of shear thickening, the polishing slurry needs to flow through the polishing area and shear relative to the polishing area, resulting in a shear effect so that the particle cluster wrapped in the abrasive particles is formed, and the workpiece is flexibly polished. The polishing of the inner arc is very easy to be interfered with by the top arc, so the influence of the installation angle on the polishing quality is particularly important. Silica sol with a particle size of 80 nm of 9 wt.% was used as the abrasive particle, the abrasive concentration the polishing speed was controlled to be 55 r/min, the rotation speed was 10 r/min, and the polishing time was 15 min. The angles of 45°, 50°, 55°, and 60° were selected for research. The influence on the surface roughness of each arc is shown in Figure 11, and the specific roughness value Ra is shown in Figure 12. When the installation angle is 55°, the inner arc polishing effect is the best, reaching Ra21.9 nm. The variation of the roughness of the outer arc and the top arc is not obvious at different installation angles.

The polishing quality becomes better with the increase in the angle because, with the increase in the angle, the polishing slurry is more likely to flow into the inner arc. Moreover, the impact force is increased, so the internal thickening agglomeration is not easy to form, forming a relatively stable flow field. However, when the angle continues to increase, the impact force exerted by the polishing slurry on the inside of the workpiece will be too large, and the internal surface will be slightly scratched.

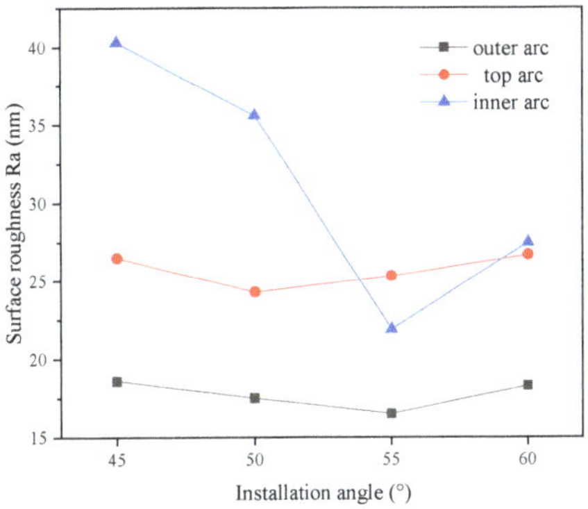

Figure 11. Effect of installation angle on arc roughness Ra.

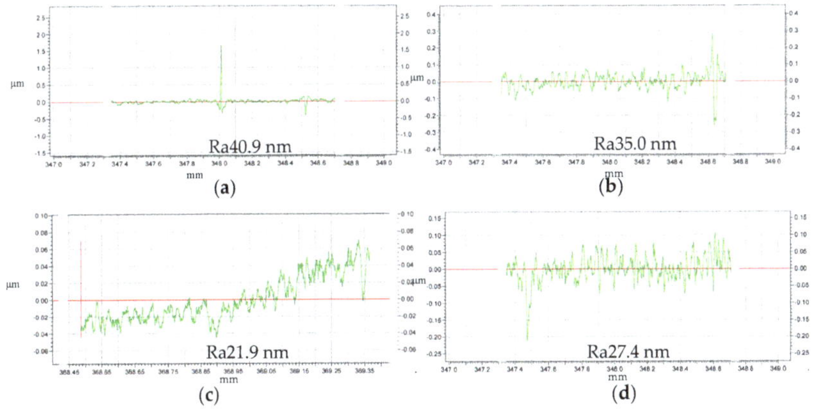

Figure 12. Measurement value of each surface roughness: (**a**) 45°; (**b**) 50°; (**c**) 55°; (**d**) 60°.

As shown in Figure 13, at a small installation angle, most of the polishing slurry directly impacts the top arc part, and only a small part of the polishing slurry enters the inner arc and is affected by the impact of the top arc and its kinetic energy decreases significantly. The decrease in kinetic energy directly affects the change in shear rate. As is known in Figure 2 above, the shear rate is positively correlated with the viscosity within a certain range, so the shear force on the inner arc will become smaller, and the polishing effect on the inner arc will also be weakened. In addition, after the inflow of polishing slurry, kinetic energy is lost again, and the fluidity is reduced again, which accelerates the formation of thickening agglomeration and further damages the uniformity of polishing. The influence of polishing slurry blockage on polishing slurry is shown in Figure 14. However, the large installation angle can make most of the polishing slurry flow directly into the inner arc so as to achieve efficient polishing, and the polishing slurry still has a certain kinetic energy after polishing. It can maintain high fluidity and make the polishing slurry continue to flow in and out. As can be seen from the figure, the large installation angle can make the entire top arc and the inner arc more balanced so that the polishing effect is more uniform.

4.4. *Effect of Polishing Speed on Surface Quality*

The polishing speed is directly manifested as the rotating speed of the polishing disc. If the rotating speed is too high, many pits will be rushed out on the surface of the workpiece, resulting in poor polishing quality. The speed is too low to reach the critical shear rate; there is no polishing effect, so the selection of polishing speed is particularly important.

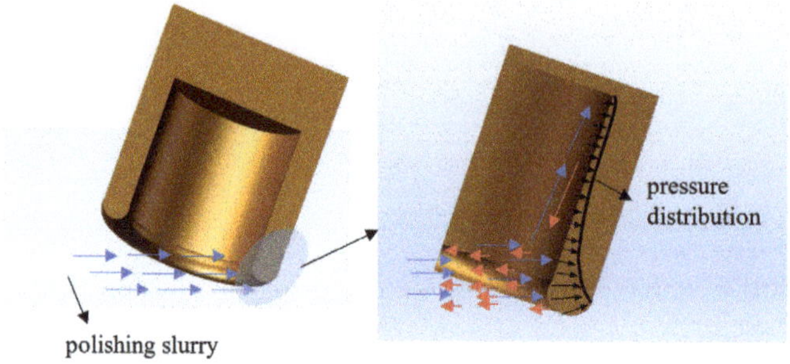

(a)

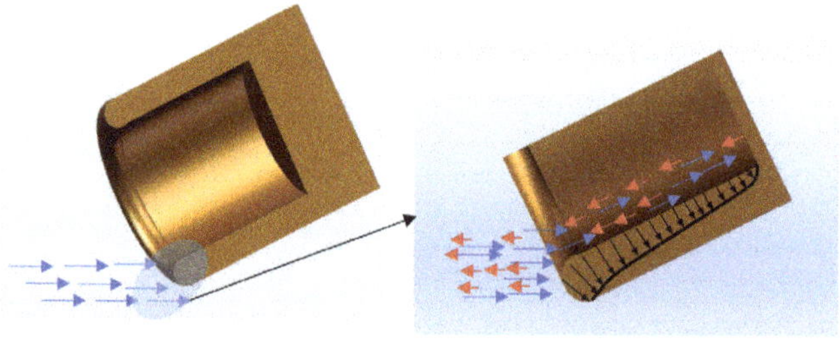

(b)

Figure 13. Influence of inclination angle on the flow state of polishing slurry: (**a**) Small inclination angle; (**b**) Large inclination angle.

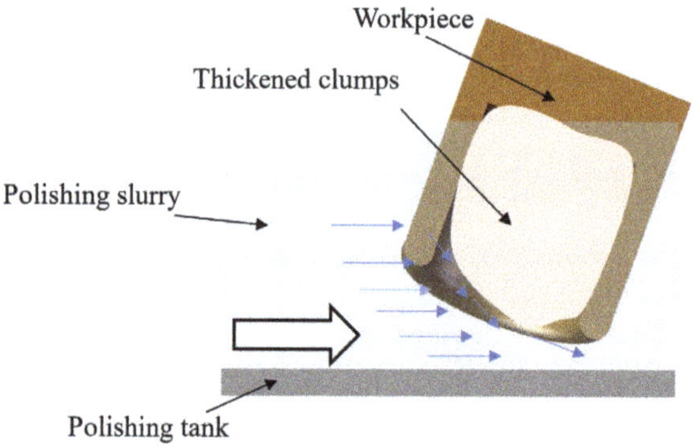

Figure 14. Schematic of thickened clumps.

The roughness variation of polishing speed at 45 r/min, 50 r/min, 55 r/min, and 60 r/min was studied. It can be found that in the inner arc, the roughness decreases with the increase in polishing speed. However, the roughness Ra of the top arc and the outer arc at 60 r/min is greater than that of 55 r/min, and the results are shown in Figure 15. The surface morphologies of each arc after polishing at different speeds are shown in Figures 16 and 17, and the specific roughness values are shown in Figure 18.

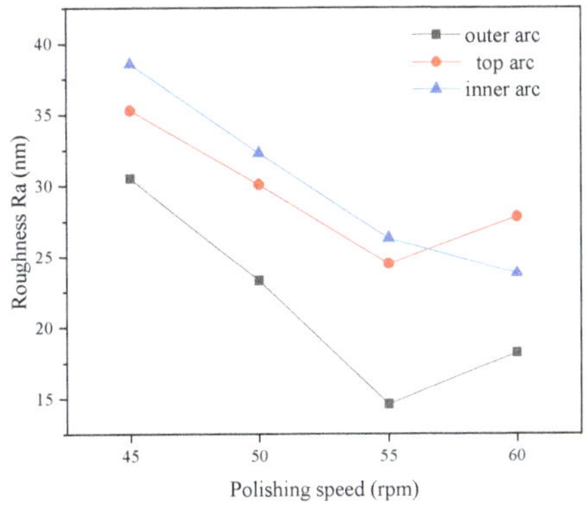

Figure 15. Effect of polishing speed on arc roughness Ra.

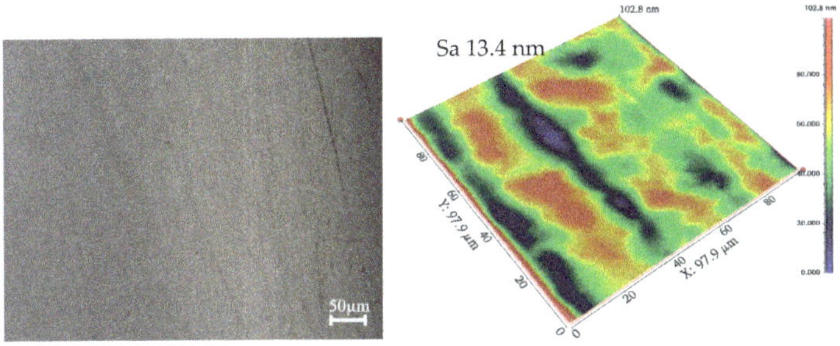

(a)

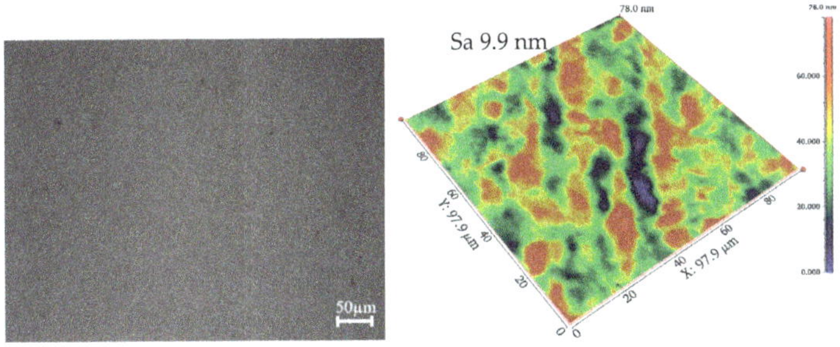

(b)

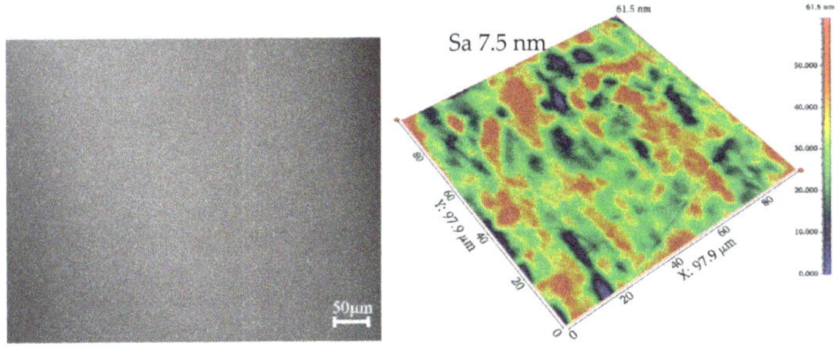

(c)

Figure 16. *Cont.*

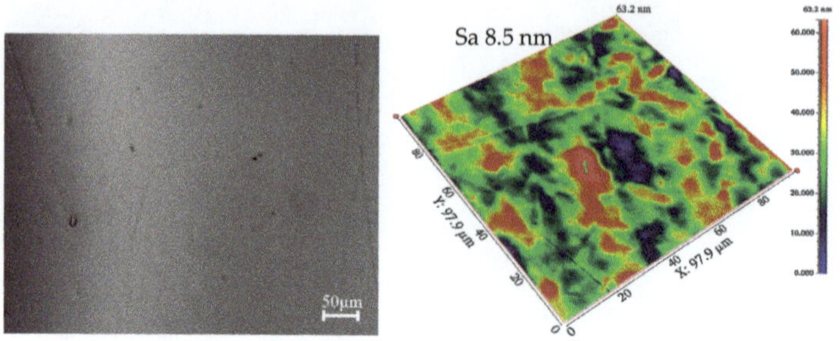

(**d**)

Figure 16. Surface topography of polishing top arc at different speeds: (**a**) 45 rpm Sa 13.4 nm; (**b**) 55 rpm Sa 9.9 nm; (**c**) 55 rpm Sa 7.5 nm; (**d**) 60 rpm Sa 8.5 nm.

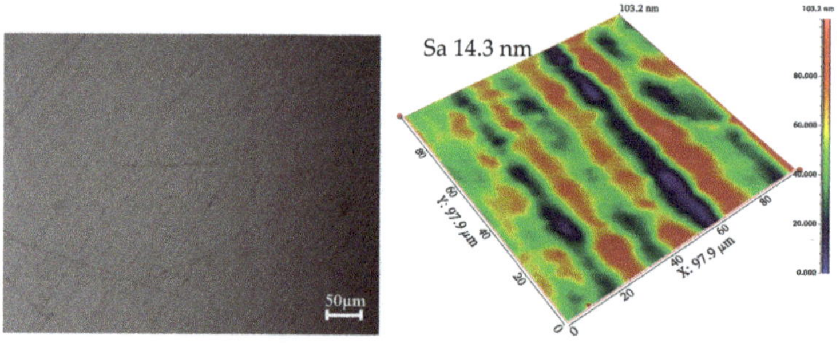

(**a**)

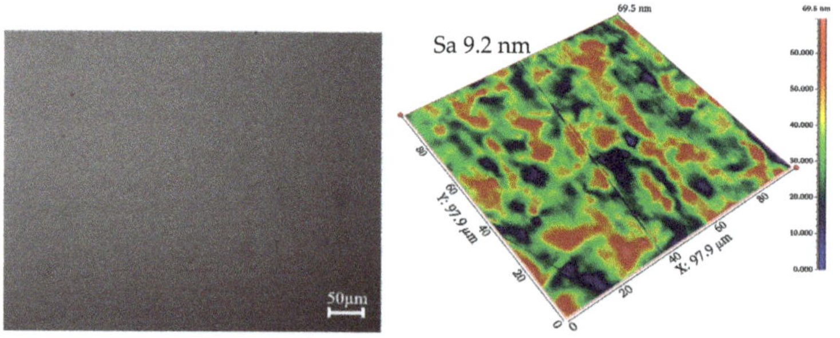

(**b**)

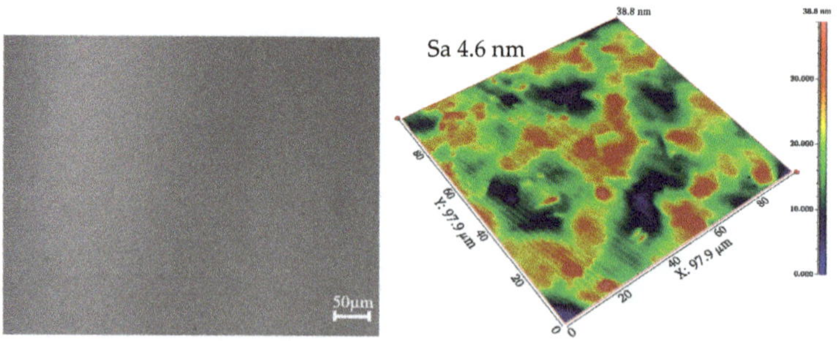

(**c**)

Figure 17. *Cont.*

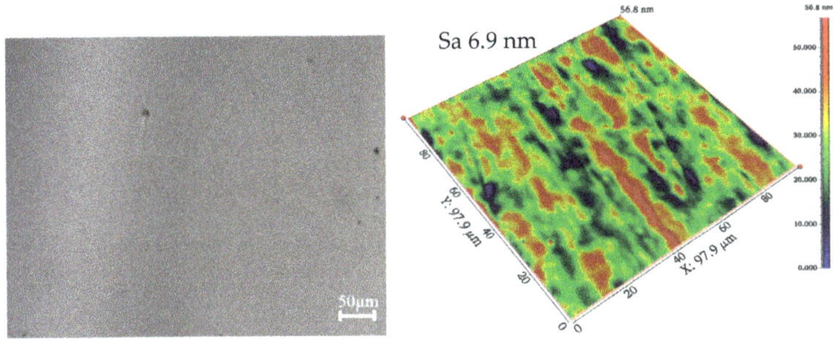

(d)

Figure 17. Surface topography of the outer arc polished at different speeds: (**a**) 45 rpm Sa 14.3 nm; (**b**) 55 rpm Sa 9.2 nm; (**c**) 55 rpm Sa 4.6 nm; (**d**) 60 rpm Sa 6.9 nm.

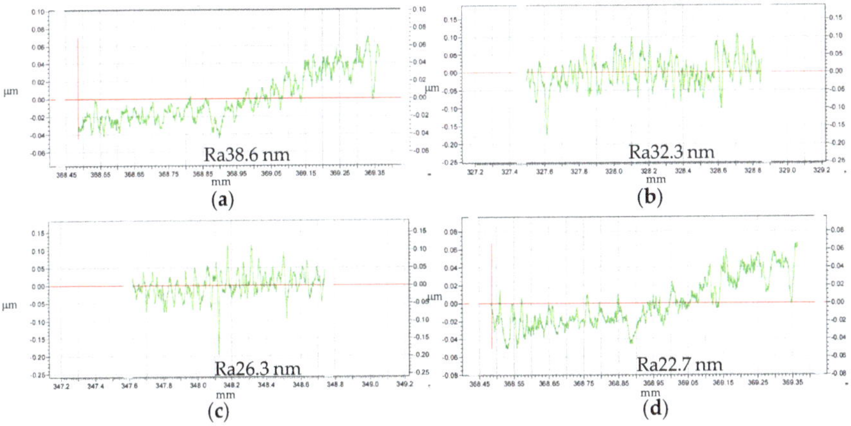

Figure 18. Surface topography of the inner arc polished at different speeds: (**a**) 45 rpm-Ra 38.6 nm; (**b**) 50 rpm-Ra 32.3 nm; (**c**) 55 rpm-Ra 26.3 nm; (**d**) 60 rpm-Ra 22.7 nm.

The top arc is the polishing area most easily in contact with the flow field. According to the rheological curve, as the rotational speed increases, the shear rate and viscosity in this area increase, resulting in an increase in shear force. The abrasive particles trapped by the shear thickening effect of colloid particles are more easily extruded into the surface of the workpiece, resulting in a large number of spot pits. The formation process of the point pits is shown in Figure 19. The surface topography of point pits is shown in Figure 20.

4.5. Selected Conditions

The combination of preferred parameters for shear thickening polishing of stainless-steel parts is shown in Table 4. The preferred parameters shown in the table are used to polish the surface of rotary workpieces with large curvature edges. The changes in surface position roughness over time during the polishing process are shown in Figure 21. After 15 min polishing, the roughness Ra of the arc section can be reduced to below 30 nm, the surface is bright, and no scratches are visible to the unaided eye. The surface morphology and roughness are shown in Figures 22–24, and the roughness results after polishing are shown in Table 5.

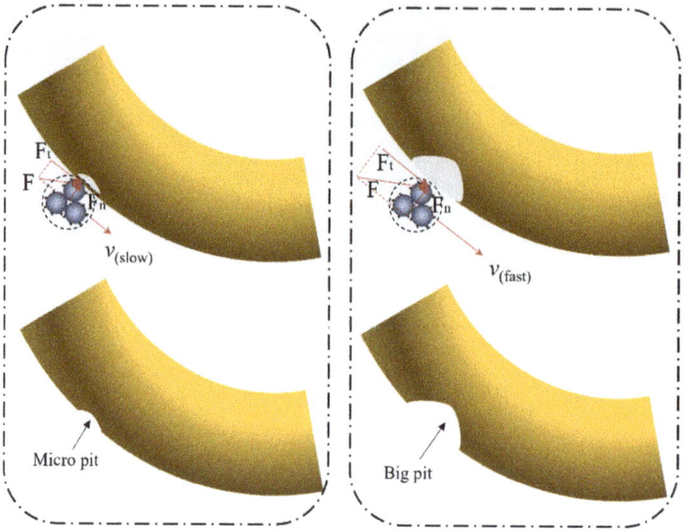

Figure 19. Schematic diagram of abrasive action at different polishing speeds.

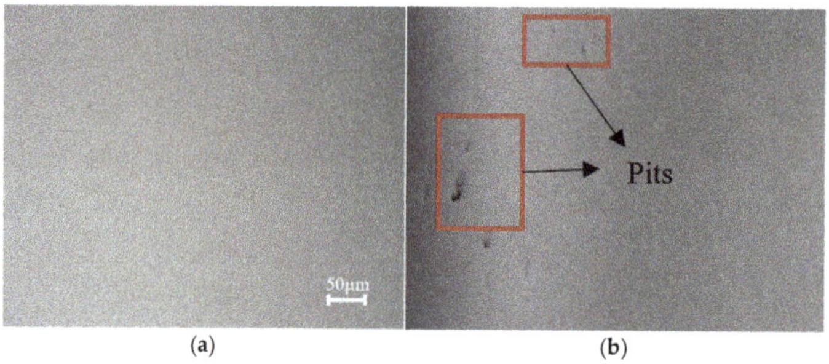

Figure 20. The surface topography of point pits: (**a**) Low polishing speed; (**b**) High polishing speed.

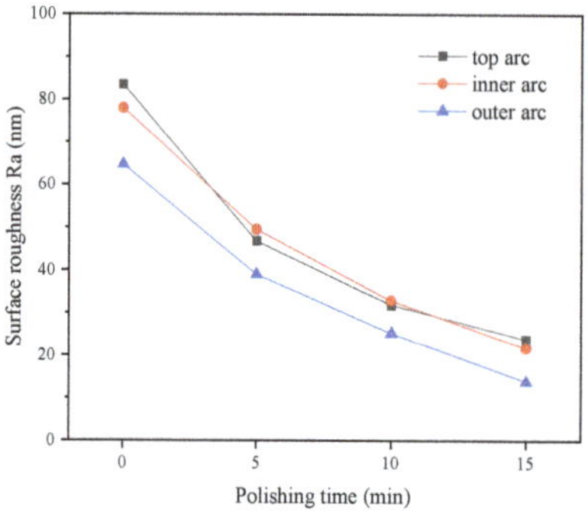

Figure 21. Variation diagram of arc roughness Ra with time after process optimization.

Table 4. Experimental parameter combination of optimized polishing.

Abrasive	Diamond Concentration	Abrasive Size	Polishing Speed	Installation Angle
Colloidal silica	9 wt.%	80 nm	55 rpm	55°

Figure 22. Macroscopic comparison of the surface morphology after polishing.

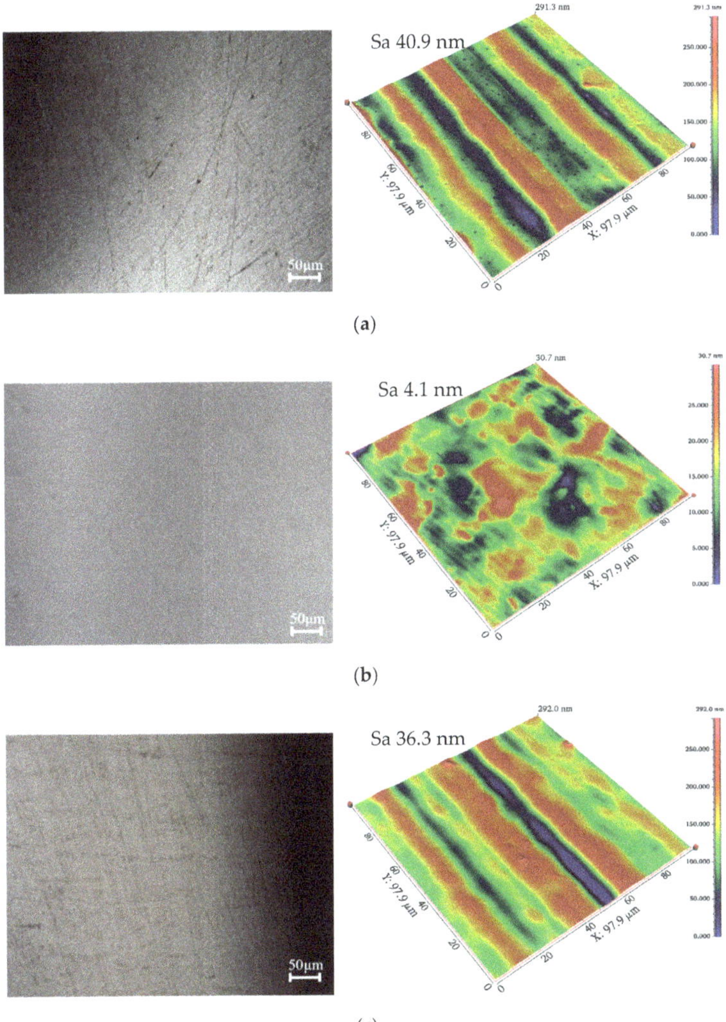

Figure 23. *Cont.*

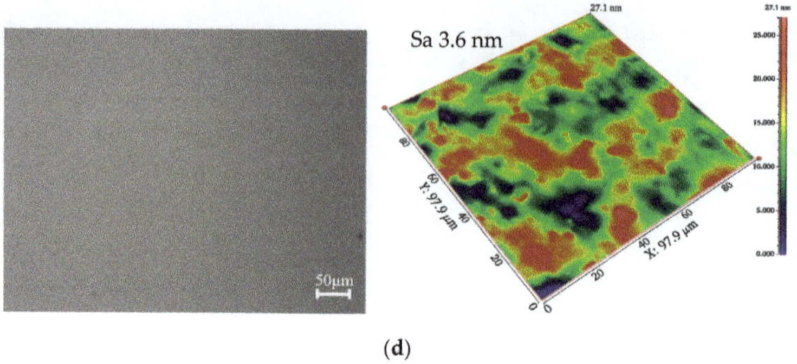

(**d**)

Figure 23. Surface topography before and after polishing: (**a**) Top arc Ra 83.4 nm/Sa 40.9 nm (before polishing); (**b**) Top arc Ra 23.8 nm/Sa 4.1 nm (after polishing); (**c**) Outer arc Ra 63.9 nm/Sa 36.3 nm (before polishing); (**d**) Outer arc Ra 14.3 nm/Sa 3.6 nm (after polishing).

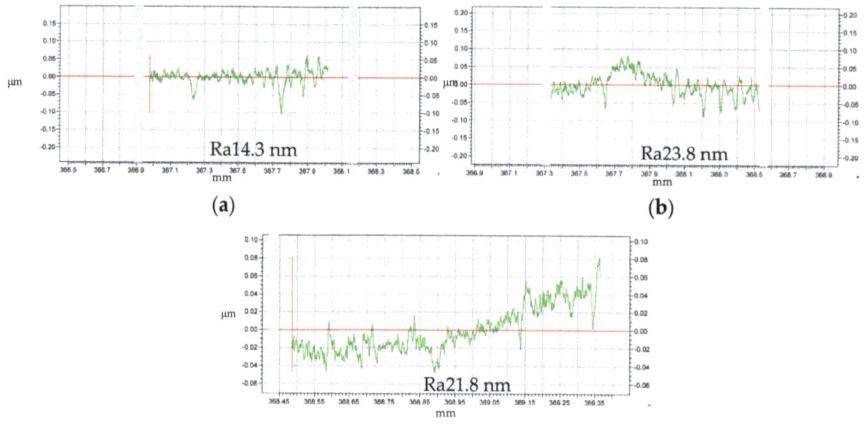

Figure 24. The roughness value of each circular arc after polishing under the optimal parameter. (**a**) Outer arc; (**b**) Top arc; (**c**) Inner arc.

Table 5. Experimental results of optimized polishing.

Position	Roughness R_a (nm)		
	Outer Arc	Top Arc	Inner Arc
Initial surface	63.9 ± 5.7/36.3 ± 2.4	83.4 ± 4.3/40.9 ± 3.1	78.0 ± 4.4
Polishing surface	14.3 ± 3.5/3.6 ± 1.8	23.8 ± 2.8/4.1 ± 1.5	21.8 ± 3.3

At the same time, traditional cashmere ball polishing was conducted on rotary workpieces with large curvature edges using the diamond paste of W5 as abrasive grains. The cashmere ball head was fixed on the hand grinder for polishing, and abrasive grains were continuously added to the polishing position during the process. The specific experimental parameters are shown in Table 6. After 15 min of continuous polishing, there were still many scratches and pits on the surface, which may be due to the unstable pressure of the hand grinder, easily causing the abrasive grains to embed into the surface of the workpiece, the surface roughness of the outer arc and top arc around 30 nm. After 30 min, the surface quality of individual areas was similar to that of the shear thickening polishing parts, but slight defects were still inevitable, with a large number of burned areas, the surface roughness of the outer arc and top arc around 20 nm. For the inner arc, and due to the influence of the size of the cashmere ball head, polishing was more difficult and the surface quality was poor. The inner arc surface roughness finalized at 34.2 nm. The specific polishing effects are shown in Figure 25.

Table 6. Experimental conditions of cashmere ball polishing.

Abrasive	Diluent Dripping Speed	Abrasive Size	Polishing Speed	Size of Cashmere Ball Head.
Diamond	4–5 drops/minute	W5	6000 rpm	20 × 3 mm

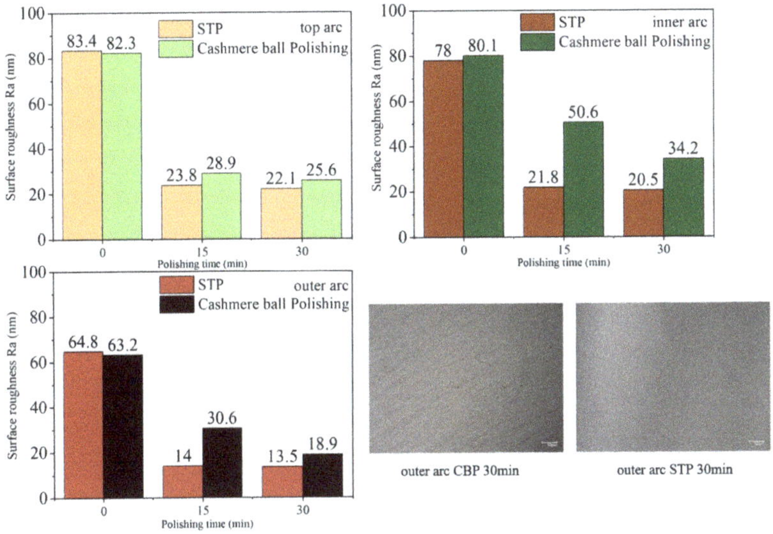

Figure 25. Comparison of results of cashmere ball polishing and shear thickening polishing.

5. Conclusions

To enhance the surface quality of rotary workpieces with large curvature edges, shear thickening polishing tests were conducted using the single factor method. The influence of various process parameters on surface roughness was systematically investigated. The key findings are summarized as follows:

(1) Abrasive Selection: High hardness and large-sized abrasives exhibit a higher material removal rate. However, they tend to cause scratches and holes during the polishing process. Silica sol, as a polishing particle, proves to be more suitable for achieving a smoother surface.

(2) Process Parameters:

- Within a certain range, increasing the abrasive particle concentration improves both polishing efficiency and quality, resulting in lower surface roughness values.
- Polishing quality improves with an increase in the installation angle. At an angle of 55°, the inner arc achieves the best polishing effect, with a surface roughness (Ra) of 21.8 nm.
- A polishing speed of 55 r/min yields optimal results for the arc section. Higher speeds, however, can lead to pit formation at the top arc position.

(3) Optimized Polishing Conditions: Polishing tests were performed under preferred parametric conditions: silica sol (80 nm) as the abrasive, an abrasive concentration of 9 wt.%, a polishing speed of 55 r/min, and a workpiece inclination angle of 55°. After 15 min of shear thickening polishing, the surface roughness of the outer arc decreased from (63.9 ± 5.7) nm to (14.3 ± 3.5) nm, the top arc decreased from (83.4 ± 4.3) nm to (23.8 ± 2.8) nm, and the inner arc decreased from (78.0 ± 4.4) nm to (21.8 ± 3.3) nm. The flow field of the polishing slurry under these conditions became more stable and effective, significantly contributing to the improved surface quality of the rotary workpiece.

(4) Comparison with Traditional Methods: In terms of both polishing quality and efficiency, the force rheological polishing method outperforms traditional cashmere ball polishing.

(5) This research validates the effectiveness of the shear thickening polishing method for rotary workpieces with large curvature edges. It provides a solid foundation and valuable insights for future polishing applications in this area.

Author Contributions: Conceptualization, Z.T. and B.L.; methodology, Z.T. and J.W.; validation, Z.S., W.Z. and L.W.; formal analysis, Z.T.; data curation, Z.T.; writing—original draft preparation, Z.T.; writing—review and editing, Z.S.; supervision, J.W and Z.S. All authors have read and agreed to the published version of the manuscript.

Funding: This research was financially supported by the National Natural Science Foundation of China (U24A20124 and U20A20293).

Institutional Review Board Statement: Not applicable.

Informed Consent Statement: Not applicable.

Data Availability Statement: Data is contained within the article.

Conflicts of Interest: The authors declare that they have no known competing financial interests or personal relationships that could have appeared to influence the work reported in this paper.

References

1. Zhou, Z. Current Research and Development Trends of Complex Surface Machining Technology. *J. Mech. Eng.* **2010**, *46*, 105–1 [CrossRef]
2. Tsay, D.M.; Chen, H.C. A Study on Five Flank Machining of Centrifugal Compressor Impellers. *J. Eng. Gas Turbines Power* **20** *124*, 177–181. [CrossRef]
3. Song, J.; Yao, Y. Material Removal Model Considering Influence of Curvature Radius in Bonnet Polishing Convex Surface. *Chin Mech. Eng.* **2015**, *28*, 1109–1116. [CrossRef]
4. Khatri, N.; Manoj, J.X.; Mishra, V. Experimental and simulation study of nanometric surface roughness generated duri magnetorheological finishing of silicon. *Mater. Today Proc.* **2018**, *5*, 6391–6400. [CrossRef]
5. Bouland, C.; Urlea, V.; Beaubier, K. Abrasive flow machining of laser powder bed-fused parts: Numerical modeling a experimental validation. *J. Mater. Process. Technol.* **2019**, *273*, 116262. [CrossRef]
6. Liu, Z.; Xiao, L.; Liu, X. Abrasive water flow polishing of superalloy microholes drilled by electrical discharge machining. *Pre Eng.* **2023**, *84*, 110–118. [CrossRef]
7. Ke, X.; Wu, W.; Shi, C. Theoretical and experimental investigation of material removal in semi-rigid bonnet polishing of binderl tungsten carbide. *J. Mater. Res. Technol.* **2023**, *24*, 1597–1611. [CrossRef]
8. Ji, S.M.; Zeng, X.; Jing, M.S. A new method for free surface polishing based on soft-consolidation abrasive Pneumatic Wheel. *A Mater. Res.* **2012**, *497*, 190–194.
9. Wang, Y.; Zhang, Y.; Feng, Z. Analyzing and improving surface texture by dual-rotation magnetorheological finishing. *Appl. Si Sci.* **2016**, *360*, 224–233. [CrossRef]
10. Li, J.; Fan, Z.; Gao, J. Multi-pole magnetorheological shear thickening polishing on inner surface of aluminum alloy slender tub *J. Mater. Res. Technol.* **2023**, *27*, 8258–8270. [CrossRef]

Liu, H.; Cheng, J.; Wang, T. Magnetorheological finishing of an irregular-shaped small-bore complex component using a small ball-end permanent-magnet polishing head. *Nanotechnol. Precis. Eng.* **2019**, *2*, 125–129. [CrossRef]
Kumar, M.; Kumar, A.; Bharti, R.K. A review on rheological properties of magnetorheological fluid for engineering components polishing. *Mater. Today Proc.* **2022**, *56*, A6–A12. [CrossRef]
Wang, A.C.; Cheng, K.C.; Chen, K.Y. A study on the abrasive gels and the application of abrasive flow machining in complex-hole polishing. *Procedia CIRP* **2018**, *68*, 523–528. [CrossRef]
Fu, Y.; Gao, H.; Yan, Q. Rheological characterisation of abrasive media and finishing behaviours in abrasive flow machining. *Int. J. Adv. Manuf. Technol.* **2020**, *107*, 3569–3580. [CrossRef]
Pan, Z.; Wan, Q.f.; Lei, Y. Research on Motion of Abrasive in Post-mixed AbrasiveWater Jet. *Mach. Tools Hydraul.* **2014**, *42*, 109–112. (In Chinese) [CrossRef]
Booij, S.M. Fluid Jet Polishing: Possibilities and Limitations of a New Fabrication Technique. Ph.D. Thesis, Technical University of Delft, Delft, The Netherlands, 2004.
Yang, R.; Lin, X. Wear behaviour of abrasive tools and its effect on surface integrity during polishing superalloy GH4169G with abrasive cloth flap wheels. *J. Manuf. Process.* **2024**, *120*, 286–304. [CrossRef]
Li, M.; Lyu, B.H.; Yuan, J.L. Shear-thickening polishing method. *Int. J. Mach. Tools Manuf.* **2015**, *94*, 88–99. [CrossRef]
Shao, Q.; Lyu, B.H.; Yuan, J.L. Shear thickening polishing of the concave surface of high-temperature nickel-based alloy turbine blade. *J. Mater. Res. Technol.* **2021**, *11*, 72–84. [CrossRef]
Wang, B.; Tang, M.; Song, C. Effects of shear thickening fluid preparation on tool performance in the milling of Ti6Al4V titanium alloy. *J. Manuf. Process.* **2024**, *113*, 95–104. [CrossRef]
Nguyen, D.N.; Dao, T.P.; Prakash, C. Machining parameter optimization in shear thickening polishing of gear surfaces. *J. Mater. Res. Technol.* **2020**, *9*, 5112–5126. [CrossRef]
Wang, J.H.; Ke, M.F. Cutting edge preparation of microdrills by shear thickening polishing for improved hole quality in electronic PCBs. *Front. Mech. Eng.* **2024**, *19*, 15. [CrossRef]
Zhou, Z.; Zhou, R.; Zhu, T. Shear-thickening effect in electrorheological fluid-assisted polishing. *Mater. Lett.* **2024**, *377*, 137342. [CrossRef]
Zhu, Z.H.; Huang, P. Fast-tool-servo-controlled shear-thickening micropolishing. *Int. J. Mach. Tools Manuf.* **2023**, *184*, 103968. [CrossRef]
Zhou, Y.; Zhang, J.; Shao, X. High Efficiency Preparation of Microdrill Edge by Shear Thickening Polishing. *Int. J. Precis. Eng. Manuf.-Green Technol.* **2024**, *12*, 75–95. [CrossRef]
Wang, J.H.; Lyu, B.H.; Jiang, L. Chemistry enhanced shear thickening polishing of Ti–6Al–4V. *Precis. Eng.* **2021**, *72*, 59–68. [CrossRef]
Chen, Y.; Jiang, M.; He, C. Microstructures and mechanical properties of dissimilar joints between cobalt-based superalloy FSX-414 and additively manufactured 316L stainless steel. *Mater. Today Commun.* **2025**, *42*, 111087. [CrossRef]
Jamalkhani, M.; Dorula, M.; Roberts, E. Densification kinetics, microstructural evolution and mechanical properties of isothermally sintered binder jetted 316L stainless steel. *J. Manuf. Process.* **2024**, *125*, 267–282. [CrossRef]

coatings

MDPI

Article

Study on the BRDF for the Surfaces of Typical Metal Materials

Zhiqiang Yang [1], Shilin Zhang [1], Yaqing Li [1], Qiang Xu [2], Liguo Wang [1] and Lei Gong [1,*]

1 School of Opto-Electronic Engineering, Xi'an Technology University, Xi'an 710021, China; yangzhiqiang@xatu.edu.cn (Z.Y.); zhangshilin@st.xatu.edu.cn (S.Z.); liyaqing@xatu.edu.cn (Y.L.); kinglywang3773@163.com (L.W.)

2 Hubei Huazhong Changjiang OP Toelectronic Technoldgy Co., Ltd., Xiaogan 432009, China; qixinyao@whcjgd.com

* Correspondence: gonglei@xatu.edu.cn

Abstract: The surface scattering property is one of the important indicators of the optical properties of metal materials, which is of great significance for the development, processing, and application of optical devices. The bidirectional reflection distribution function (BRDF), as an optical mathematical model, describes the spatial distribution of reflected light energy on the surface of an object and can provide an important basis for the characterization of the optical properties of the surface of an object. In this paper, the white light scattering characteristics of the surfaces of four metallic materials, namely brass, aerospace aluminum, oxygen-free copper, and red copper, are investigated based on the BRDF microfacet metamodel, and the scattered field distributions are explored for different angles of incidence, scattering angles, and azimuthal angles. This study provides a reference for the evaluation of the optical properties of metal material surfaces and provides a theoretical basis and technical support for the design and application of optical devices.

Keywords: BRDF; typical metal material; surface reflection; scattering properties

1. Introduction

In the context of the accelerated advancement in the domain of aerospace science and technology, there has been an emergence of increasingly stringent requirements for the suppression of stray light in critical-space low-light detection systems, such as satellite laser communication and gravitational wave detection. The surface scattering characteristics of the optical components directly determine the distribution of stray light within the system. Non-target stray light outside the field of view, caused by light scattering from the surface of the metal structural components, severely restricts the signal-to-noise ratio of the system, which greatly affects the stability of the system.

Brass has excellent mechanical properties, physical properties, and high electrical and thermal conductivity and corrosion resistance; it is widely used in electronics, telecommunications equipment, household appliances, connectors, and accessories. Red copper exhibits high electrical conductivity and good wear resistance, making it a suitable material for use in conductive springs, wires, and cables. These characteristics position the copper alloy for a broad range of applications in the aviation sector, where stringent material performance requirements are in place. The bidirectional reflectance distribution function (BRDF) is an effective means to characterize the scattering properties of target surfaces. In this study, a microelement model is employed to derive expressions for the contributions of both specular and diffuse reflection components. Experimental measurements and analyses were conducted in conjunction with the model, enabling the distillation of the governing laws underlying multifactorial influences. These aspects constitute the innovative contributions of this research.

In the field of optical characterization and materials science, the scattering properties of metal surfaces are one of the current research hotspots. The bidirectional reflection distribution function (BRDF), as a physical parameter describing light scattering properties, can be used to characterize the anisotropic scattering properties in the hemispherical space on a material surface. The study of the BRDF properties of material surfaces not only helps to understand the optical properties of material surfaces but also provides theoretical support and data support for material surface processing and practical engineering applications.

As early as 1978, Ishibashi et al. [1] conducted durability, heat, and weathering tests on selectively absorbing surfaces of stainless steel, measured specular reflection, diffuse reflection, and angular dependence and determined the absorbance of stainless steel to be 0.91–0.93 using an integrating sphere, exploring the possibility of obtaining a lower radiative property of stainless-steel surfaces in the infrared region without significantly reducing the high absorbance. Weaver et al. [2] explored the optical properties of metallic materials, which provided a reference for the experimental study of the light reflection behavior of metal surfaces. Nayar et al. [3] investigated the reflectance models based on physical and geometrical optics, and through the reflectance curves predicted by the two models, they proposed a reflectance framework consisting of three parts, the diffuse reflection flap, specular reflection flap, and specular cusp, and analyzed the effect of surface roughness on it. Ward [4] proposed a mathematical model based on the anisotropic reflectance of metal surfaces, which can effectively restore and reproduce surface spatial reflectance data.

Earlier this century, Li and Torrance [5] investigated the angular distribution of light scattered on several rough metal surfaces and used the BRDF model based on the Kirchhoff approximation to establish the relationship between the surface-height statistics and the angular distribution of the scattered light. The fitting yielded roughness statistics that matched very well with the measurements of the high-frequency small-scale roughness components. Ngan et al. [6] built a new setup for the efficient acquisition of anisotropic BRDFs, which is capable of acquiring anisotropic materials at a high resolution so that the sampling microfacet distributions computed from the measurements can be fitted with improved fit and intrinsically reproduce the measurements. Xie et al. [7] combined the effects on the surface profile from the surface wettability theory and wet conditions, conducted an experimental comparison, and measured the BRDF value of the surface of three exterior decorative materials for buildings, namely granite, coated glass, and an Ethylene Propylene Diene Monomer (EPDM), under the conditions of two dry and wet wavelength bands (visible band 0.6328 μm and near-infrared band 1.34 μm) and different angles of incidence. The data showed that if a stable water film with a translucent smooth interface function is formed above the surface of the material, its BRDF value will show obvious specular reflection components in different degrees. eZhang et al. [8] measured the BRDF at a wavelength of 660 nm for a copper foam sheet with a pore size of 90 ppi and a porosity of 98.6% at different incidence angles of 0°, 15°, 30°, 45°, 60°, and 75°, respectively, and the results showed that the reflected energy at the backward direction is greater than the forward in the conditions of oblique incidence; with the increase in the incident angle, the asymmetry parameter decreases, the high reflected energy area becomes larger, and the integrated hemispherical reflectance increases. Vitali et al. [9] presented Vis-NIR spectral BRDF measurements of laser-polished metallic, opaque, flat samples commonly employed in laser metal machining and concluded that in addition to the roughness, the optical properties seem to depend on the atmospheric composition of the polishing process.

Renhorn et al. [10] built a physics-based model for a rough-surface BRDF that accurately describes two orders of magnitude of BRDF scattering data for easy numerical computation, taking into account angles of incidence and scattering, the effective index, surface autocovariance, and correlation length. Durikovic et al. [11] conducted BRDF spectral measurements, rendered synthetic images of metallic paints, measured the spectral reflectance of multiple paint samples, and used the measured data to fit the analytical BRDF model to obtain the directional optical effect of incident radiation parameter. Cui et al. [12] explored the relationship between the incident angle, surface temperature, and scattering characteristics of metal surfaces and showed that the temperature field has a more pronounced effect on the scattering characteristics.

In recent years, researchers and scholars have begun to focus on the application and optimization of BRDF in specific scenarios, such as the study of BRDF optimization modeling for painted metal materials [13]. Ma et al. [14] conducted spectrally polarized BRDF measurements of brass in the visible light band, built an exponential model of the BRDF, and discussed the influences of wavelength, angle, polarization of incident, and reflected light on the BRDF, showing that the effect of wavelength on the BRDF depends on the color of the sample and that the polarization of incident light has little influence on the variation tendency of the BRDF versus the incident zenith angle and wavelength. Zou et al. [15] reviewed the development of devices to measure the BRDF, collected research results in this field, summarized outstanding problems related to BRDF measurement, and proposed directions of future research in the area. Chen et al. [16] measured the BRDF in the near-infrared band of brass rough surfaces under variable temperature conditions and built a six-parameter model of the BRDF, which showed that the temperature, heating time, and surface roughness have significant effects on the BRDF and that the overall error is not higher than 3.29%. Research on the BRDF of rough metal surfaces and deterministic structures [17,18] has been investigated by several early scholars. Miao Yupei et al. [19] established a hybrid polarization model for metal surfaces that integrates both specular reflection and inter-reflection based on the physical mechanism of Fresnel reflection between light and metal surfaces. By employing microfacet theory, they analyzed the specular reflection BRDF while approximating the inter-reflection BRDF using the phase Mueller matrix, ultimately deriving the hybrid reflection loss for multi-view 3D reconstruction. Xie Yinmo et al. [20] conducted a comparative analysis of the BRDF for anisotropic machined surfaces produced through three different processing methods (face milling, planing, and surface grinding) under varying incident angles, while also measuring and analyzing the directional reflection characteristics of fabrics with distinct texture structures.

This paper builds a BRDF measurement system. The overall system consists of four subsystems: light source, scanning, detection, and software control. It realizes operation control and scanning trajectory parameter setting, sets different experimental variables, collects BRDF values on the surface of metal structural components, analyzes the BRDF peak and the scattering distribution law under various influencing factors, and studies the surface scattering characteristics of the components. Thereby, precise measurement of the scattering signal at the target position is achieved, which is one of the innovations of this work. In addition, the establishment of the BRDF model can fully reflect the reflection information on the surface of the metal structural components. How to establish the model based on the distribution characteristics of its microstructure and how to solve the model parameters are also the innovations of this topic.

This paper establishes a high-light BRDF model based on the microfacet element theory and analyzes and compares the scattering characteristics of the surfaces of typical metal materials, such as brass, aerospace aluminum, oxygen-free copper, and red copper, using a compact mechanized optical system. This study explores the influence of the optical properties of the material surfaces on the light scattering characteristics. Furthermore, this paper combines data related to the roughness of the material surfaces with the aforementioned data and compares the scattering characteristics of the four typical metal materials under white-light irradiation.

2. Bidirectional Reflection Distribution Function and Microfacet Theory

The bidirectional reflection distribution function is a mathematical model that describes the reflection properties of an object's surface to light and is an effective tool for understanding and modeling the interaction of light with material surfaces [21–27]. The Microfacet Element Theory is a physical model that explains the BRDF based on the light scattering properties of the surface microstructure, which assumes that the surface consists of a number of tiny mirrors (microfacet elements), each of which reflects light independently.

When the light strikes the surface along the l direction and is observed in the v direction, since the microfacet element will only reflect the incident light to the direction symmetric about the normal, only the microfacet element with the normal vector h will reflect the incident light in the l direction to the v direction, as shown in Figure 1 [28–32]:

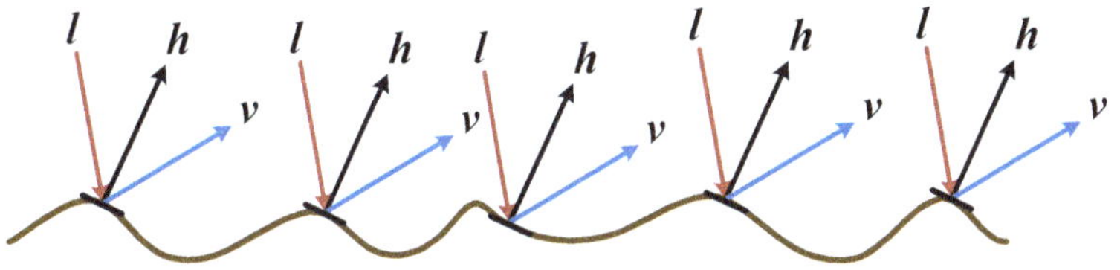

Figure 1. Schematic of light reflected from the surface of the material.

The normal distribution function (NDF) $D(h)$ is used to describe the probability of the normal distribution of all the microfacial elements that make up the surface: input a vector h to the NDF, and the NDF will return the ratio of the number of microfacial elements whose normal vector is h to the total number of microfacial elements.

When the object surface reflection occurs, due to the existence of rough surface undulations, not all microfacets can complete the reflection. As shown in Figure 2a, part of the incident light was blocked and failed to reach the surface element; this phenomenon is called the shadowing effect. As shown in Figure 2b, part of the reflected light that was blocked failed to reach the observation point; this phenomenon is called the masking effect. There are also multiple reflections of light between microfacets, as shown in Figure 2c, which may be the source of part of the diffuse light.

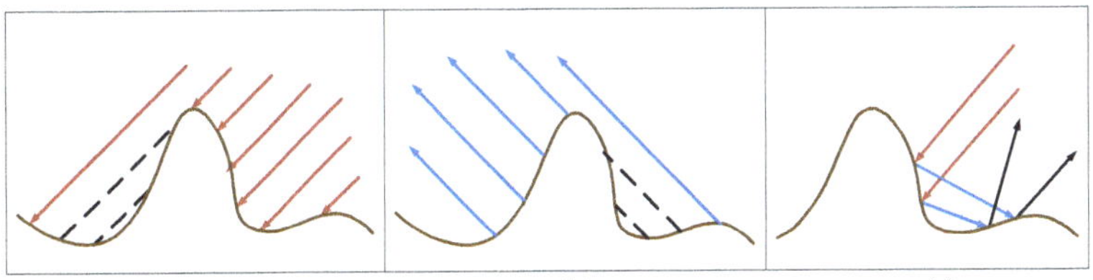

(a) (b) (c)

Figure 2. Schematic of light reflection from a microfacet.

In order to mitigate the effects of masking and shading effects of light rays between surface microfacets, a geometric attenuation factor (GAF) is introduced. In this paper, we use the geometric attenuation factor $G(l,v)$ correction model of the random surface microfacets theory [33] to calculate the value of the proportion of light that can be reflected from the l direction to the v direction without being shaded.

The light in the two different media on the critical surface will not be all reflected but will be part of the refraction; this is the proportion of the reflection in line with the Fresnel Equations $F(l,v)$.

Using the Torrance–Sparrow model [34,35], based on the microfacet element theory, which describes the reflection properties of light on rough surfaces, the behavior of light on the surface of real objects can be predicted and simulated more accurately than other models. In this paper, this model is used to model the bidirectional reflection distribution function of high light by combining the three functions mentioned above:

$$f(l,v) = \frac{F(l,h)G(l,v)D(h)}{4\cos\theta_i \cos\theta_s} = \frac{F(l,h)G(l,v)D(h)}{4(n\cdot l)(n\cdot v)} = \frac{I(S)}{4\cos\theta_i \cos\theta_s} \tag{1}$$

wherein n is the macroscopic surface normal, h is the microfacet normal, θ_i is the angle between the incident light and the macroscopic surface normal, θ_s is the angle between the reflected light and the macroscopic surface normal, and $I(S)$ represents the scattering intensity. This represents a well-established modeling framework, as evidenced by previous studies that have employed this approach with the microfacet tangent plane approximation theory. Through this methodology, scholars have derived analytical expressions quantifying the contributions of both specular and diffuse reflection components from sea surfaces, ultimately establishing a numerical model for oil film reflectance characterization on rough marine surfaces [36].

3. Experimental Methods

The instrumentation used in this experiment is a compact mechanized optical system REFLET for the scattering characterization of all materials manufactured by Light Tech, France, as shown in Figure 3. The illumination part of the instrument adopts a 100 W halogen lamp light source box, the spot size on the sample surface can be manually adjusted (Φ 1.0–13.0 mm), and the angle of the beam aperture can be manually adjusted ($\pm 0.15 - \pm 2.26^\circ$). The goniometer can be rotated from 0–180°, the resolution of the selected angle is 1°, and the positioning accuracy is 0.3°. The instrument detection part of the silicon photodiode detector has a high-precision full-band dynamic detection range of 10^9, a low-precision dynamic detection range of 10^4, and a shading angle of 4°. The angle reception range of the optical system of the instrument is $\pm 0.04^\circ$, and the size of the observation area is Φ 6 mm. The incidence angle θ_i of the goniometer ranges from -90° to 90°, and the scattering angle θ_s ranges from -90° to 90°, with an angular resolution of 0.1° and a positioning accuracy of 0.5°. The polarizer is set to 0°, and polarization is not considered. In our experiment, the white light source is used, which contains all wavelengths from 400 nm to 800 nm, and the detection part has been adjusted so that the signal is proportional to the total power of the white light.

The mechanical rotation of the REFLET measuring instrument is shown schematically in Figure 3. The measured sample is rotated on the sample holder, and for each irradiation angle θ_i of the light source for the irradiated sample, the photoelectric receiver can be driven by the mechanical rotating arm to realize the half-space scanning reception of the scattered light. The pitch angle is $0^\circ \le \theta_s \le 180^\circ$, and the azimuth angle is $0^\circ \le \phi_s \le 180^\circ$ receiving rotation. The mechanical rotary system adopts a servo motor system, and the rotation angle precision is 0.5°.

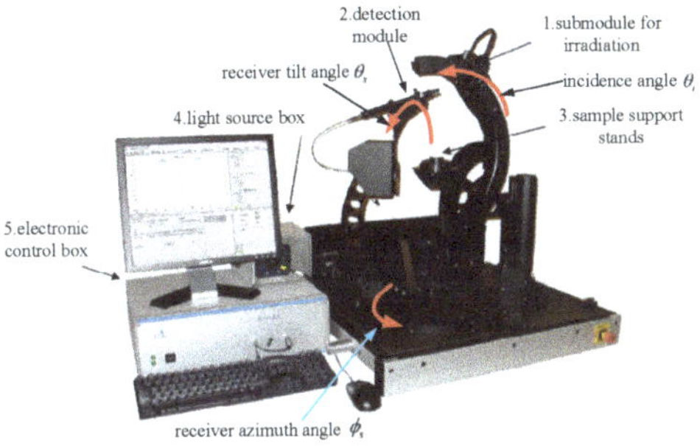

Figure 3. The REFLET system.

4. Results and Analyses

Before studying the white-light BRDF and scattering intensity of the four metal samples selected in this paper, the surface roughness of the four metal samples needed to be examined in order to determine the subsequent experiments to be conducted under the condition that the surface roughness is roughly the same, and the results shown in Figure 4 were obtained after measuring the samples using the white-light interferometer.

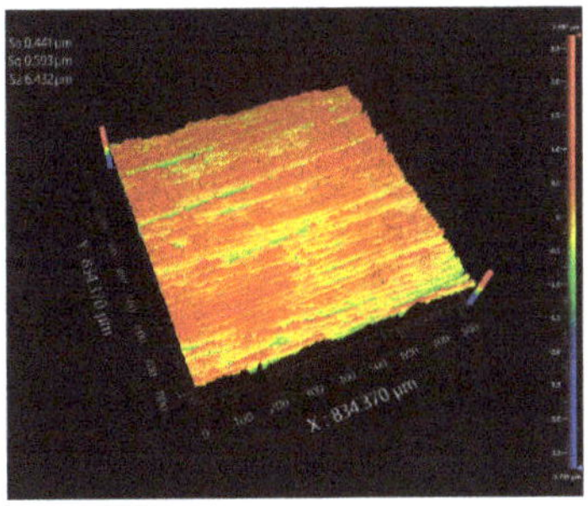

(**a**) Brass

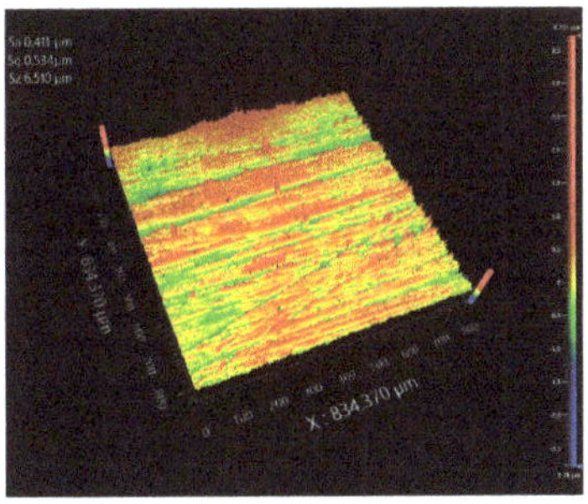

(**b**) Aerospace aluminum

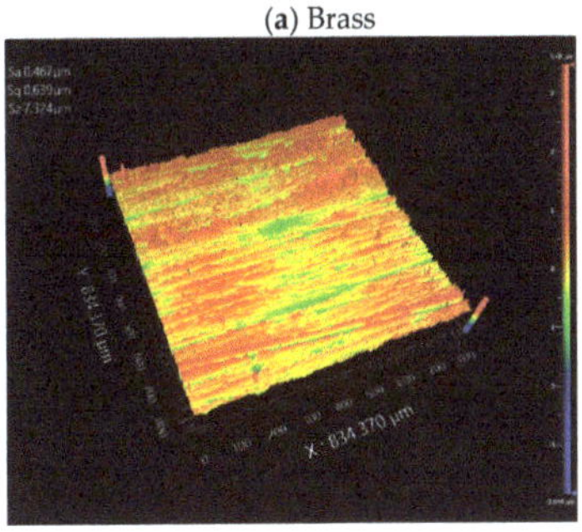

(**c**) Oxygen-free copper

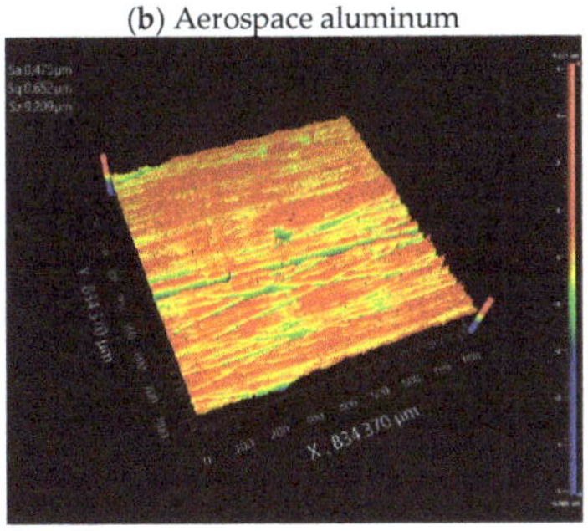

(**d**) Red copper

Figure 4. Surface roughness data for four materials.

The surface roughness data of the samples selected in this paper are displayed in Figure 4. In this figure, 'Sa' is the arithmetic mean height, indicating the arithmetic mean height of the surface, i.e., the average of the absolute values of the heights of all the points on the surface; 'Sq' is the root-mean-square height, indicating the root-mean-square height of the surface, i.e., the square root of the average of the squares of the heights of all the points on the surface; and 'Sz' is the maximum height, indicating the maximum difference in the heights of the surface, i.e., the vertical distance between the highest point on the surface and the vertical distance between the highest and lowest points on the surface. In general, 'Sa' is used to reflect the average roughness of the surface; the larger the value, the rougher the surface. The average surface roughness of the brass samples is 0.441 μm, the average surface roughness of the aerospace aluminum samples is 0.411 μm, the average surface roughness of the oxygen-free copper samples is 0.467 μm, the average surface roughness of the violet copper samples is 0.475 μm, and the surface roughness data of the four samples tested have an error in the range of 0.050 μm.

The power spectral density data for the four materials selected for this paper are shown in Figure 5, which demonstrates the power spectral density in the *X*-directions and *Y*-directions of the test surfaces of the four materials.

White light scattering occurs when it strikes the surface of a material, and the scattering intensity is affected by the angle of incidence, scattering angle, and azimuth angle. The change in the angle of incidence affects the path length and depth of the interaction between the beam and the material, thus affecting the scattering intensity.

Figures 6 and 7 show the variation curves of the white-light BRDF and scattering intensity with the scattering angle for brass at different incidence and azimuth angles. The BRDF shows how bright the surface is viewed along the scattering direction, and the scattering intensity means the total power of the scattered light along the scattering direction. It can be seen that surface is brightest around the mirroring direction, and it is brighter while the incidence angle is larger. However, the peak value of the scattering intensity has a minimum value at an incidence angle between 30° and 60°.

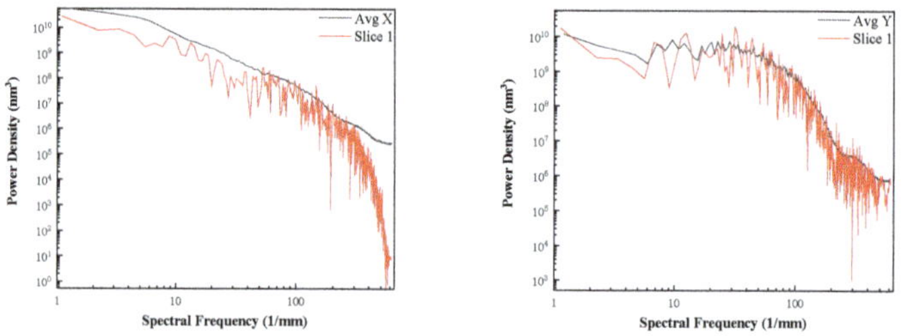

(**a**) Brass *X*-direction and *Y*-direction

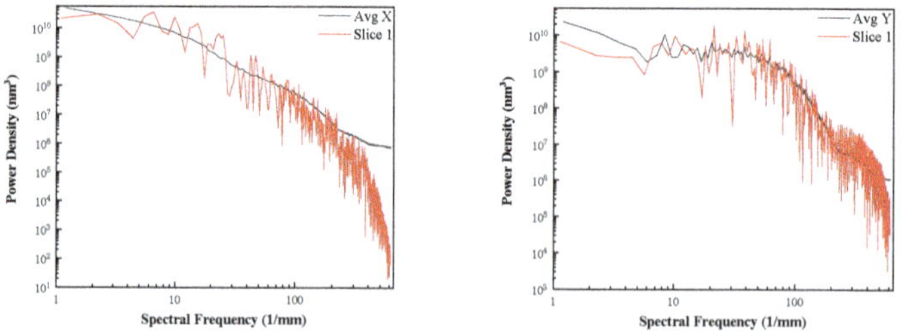

(**b**) Aerospace aluminum *X*-direction and *Y*-direction

Figure 5. *Cont.*

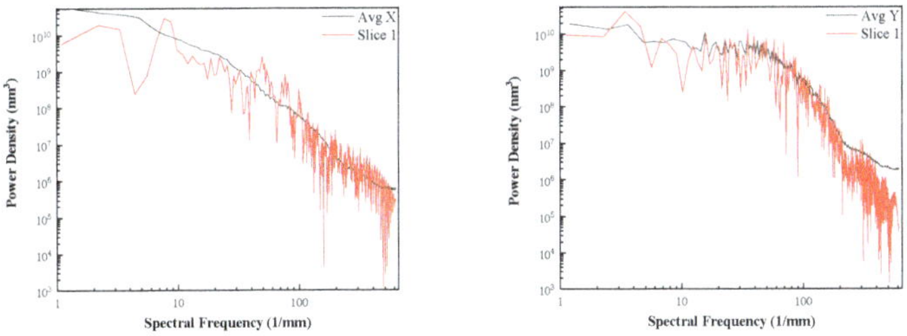

(**c**) Oxygen-free copper *X*-direction and *Y*-direction

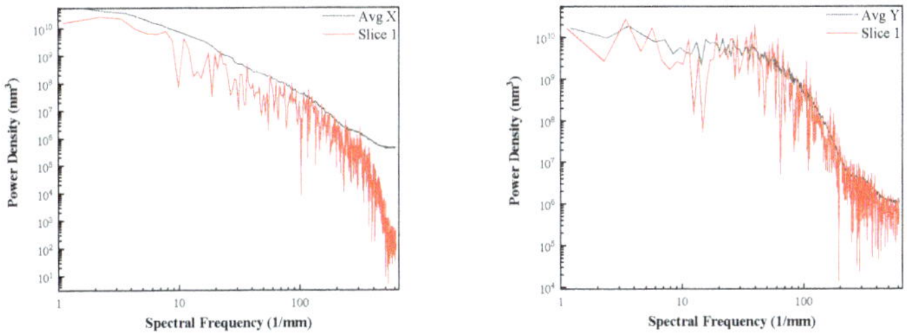

(**d**) Red copper *X*-direction and *Y*-direction

Figure 5. Power spectral density data for four materials.

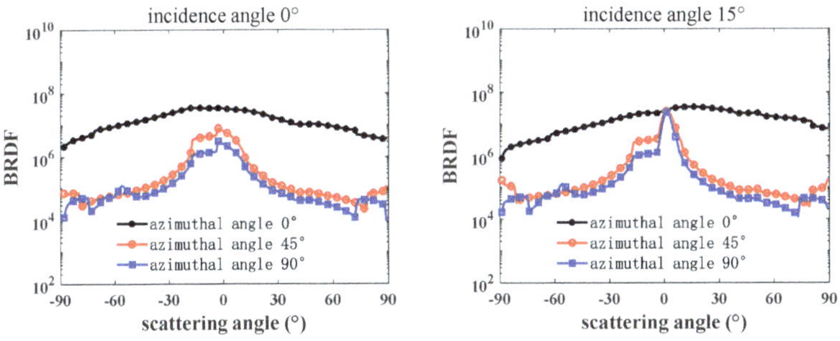

(**a**) Incidence angle 0° (**b**) Incidence angle 15°

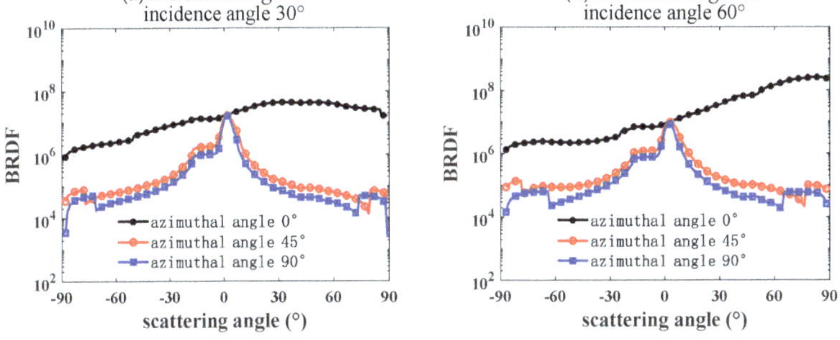

(**c**) Incidence angle 30° (**d**) Incidence angle 60°

Figure 6. Variation curves of white-light BRDF with scattering angle for brass at different incidence and azimuth angles.

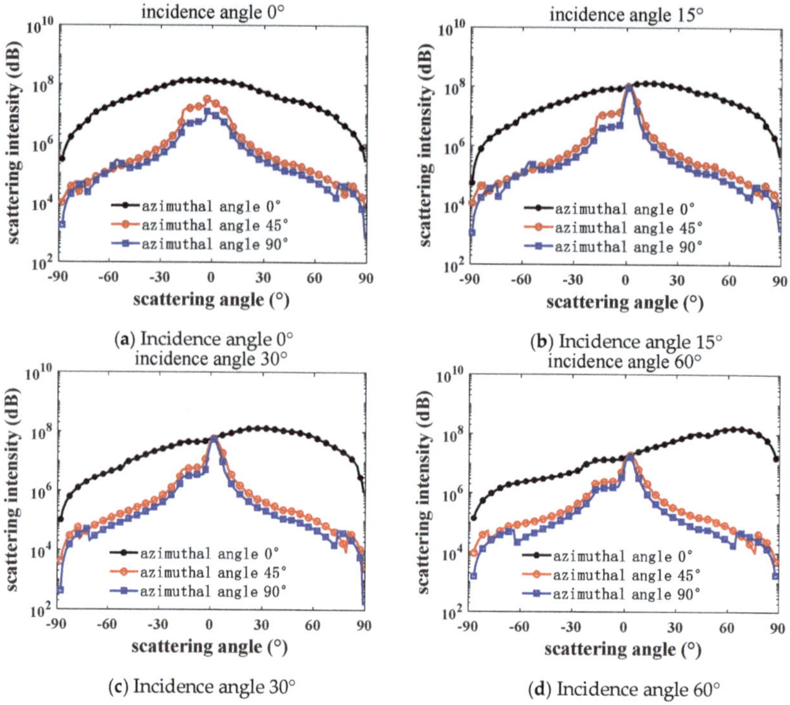

(**a**) Incidence angle 0°

(**b**) Incidence angle 15°

(**c**) Incidence angle 30°

(**d**) Incidence angle 60°

Figure 7. Variation curves of white light scattering intensity with scattering angle for brass at different incidence and azimuth angles.

Comparing and analyzing Figures 6 and 7, in the direction of the incidence angle θ_i = 0 and scattering angle θ_s = 0, the differences in the BRDF and scattering intensity at different azimuth angles in the direction of incidence angle and scattering angle are obvious, and it can be obtained that the selected brass surface is anisotropic. When the azimuth angle ϕ_s = 0°, the BRDF and scattering intensities show a tendency of the forward scattering gradually exceeding the backward scattering with the increase in the incidence angle, especially in the direction of the specular reflection of forward scattering, i.e., when $\theta_s \approx \theta_i$, the BRDF and scattering intensities reach the peak. This phenomenon indicates that the brass surface has significant specular reflection properties and good flatness.

In Figure 6, the forward scattering BRDF values are positively correlated with the incidence angle in the range of the scattering angle $\theta_s > 75°$, and the closer the scattering angle is to 90°, the more obvious it is that the BRDF value is greater than the scattering intensity value. In Figure 6, the BRDF and scattering intensity are compared at a large scattering angle $\theta_s = \pm 83°$ when the azimuth angle ϕ_s = 90° and the incident angle $\theta_i \neq 0°$. Both the forward and backward scattering intensities decrease with increasing incidence angle, which is due to the fact that at larger incidence angles, the light rays may undergo more internal reflections and absorptions from the surface, and some of the light rays are either completely absorbed or lose most of their energy after multiple reflections, resulting in a decrease in the scattering intensity.

Therefore, when the scattered light is perpendicular to the brass surface and the azimuth angle is 0°, the forward scattering has the strongest scattering intensity in the specular reflection direction. Figure 8 shows a schematic diagram of white light scattering on the brass surface, and combined with the scattering intensity curve analyzed in Figure 7, when the azimuth angle $\phi_s \neq 0°$, the scattering angle has strong scattering at around 0°. By analyzing the BRDF and surface roughness, the brass surface roughness is low, exhibiting specular high-light and directional range reflections.

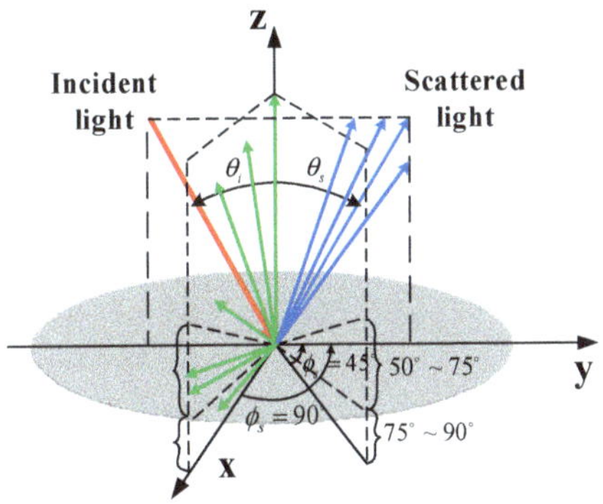

Figure 8. Schematic of white light scattering on brass surface.

Figures 9 and 10 show the variation curves of the white-light BRDF and scattering intensity with the scattering angle for aerospace aluminum at different incidence and azimuth angles.

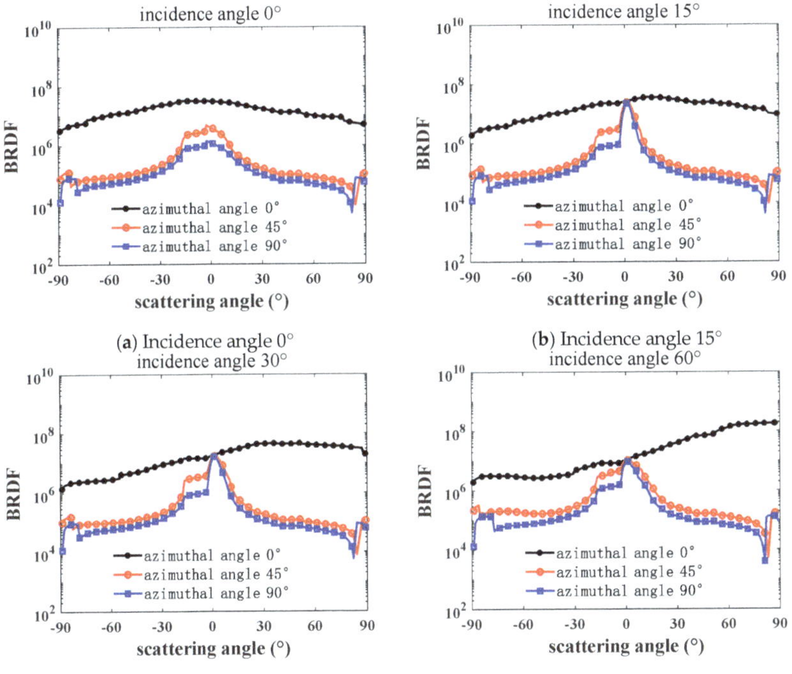

(**a**) Incidence angle 0°
(**b**) Incidence angle 15°
(**c**) Incidence angle 30°
(**d**) Incidence angle 60°

Figure 9. Variation curves of white-light BRDF with scattering angle for aerospace aluminum at different incidence and azimuth angles.

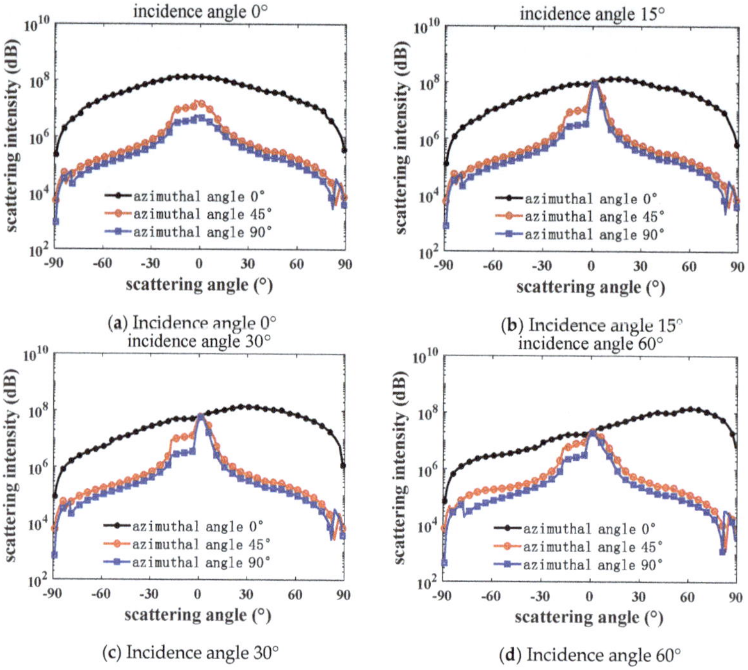

(**a**) Incidence angle 0°
(**b**) Incidence angle 15°
(**c**) Incidence angle 30°
(**d**) Incidence angle 60°

Figure 10. Variation curves of white light scattering intensity with scattering angle for aerospace aluminum at different incidence and azimuth angles.

Comparing and analyzing Figures 9 and 10, when the azimuthal angle is $\phi_s \neq 0°$, the BRDF and scattering intensity are the largest in the specular reflection direction of forward scattering (i.e., at $\theta_s \approx \theta_i$). When the azimuth angle $\phi_s \neq 0°$, the scattering angle is within 78–90°, and the BRDF and scattering intensity increase and then decrease with the increasing angle, with fluctuations in a small range. Compared with the brass surface, the BRDF and scattering intensity of the aerospace aluminum surface are much larger than those of the brass surface in the range of large scattering angles, i.e., it has better specular high-light performance than brass in the range of large scattering angles, and its directional strong scattering has a smaller range, as shown in Figure 11. Analyzing the surface roughness of aerospace aluminum, although it has a high surface roughness, its excellent high-light properties at large scattering angles make it widely used in aerospace devices such as aircraft, rockets, and satellites, where its high reflectivity and optical properties improve the performance of the device in extreme environments.

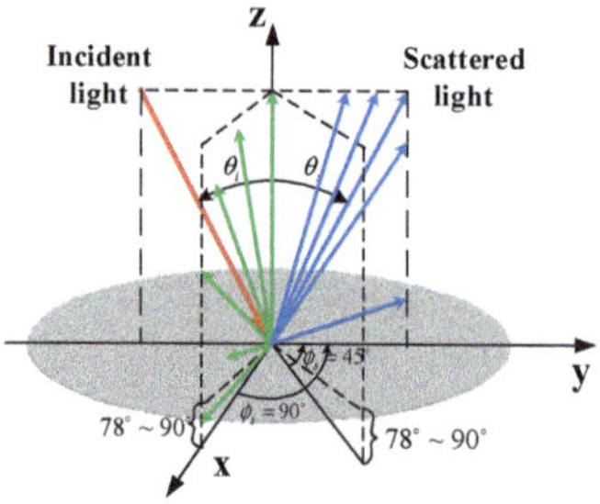

Figure 11. Schematic of white light scattering on aerospace aluminum surface.

Figures 12 and 13 show the variation curves of the white-light BRDF and scattering intensity with the scattering angle for oxygen-free copper at different incidence and azimuth angles.

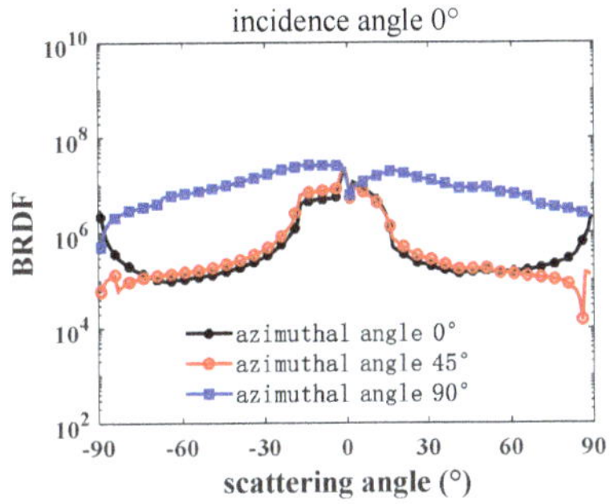

(**a**) Incidence angle 0°

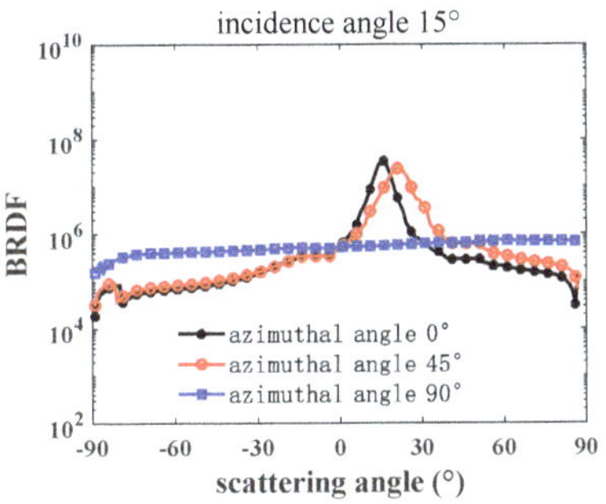

(**b**) Incidence angle 15°

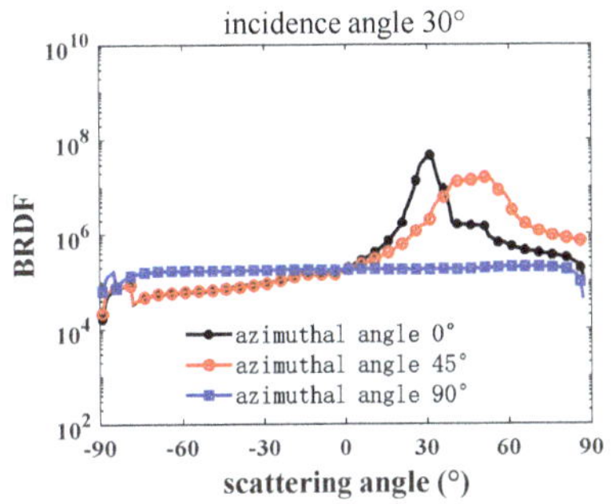

(**c**) Incidence angle 30°

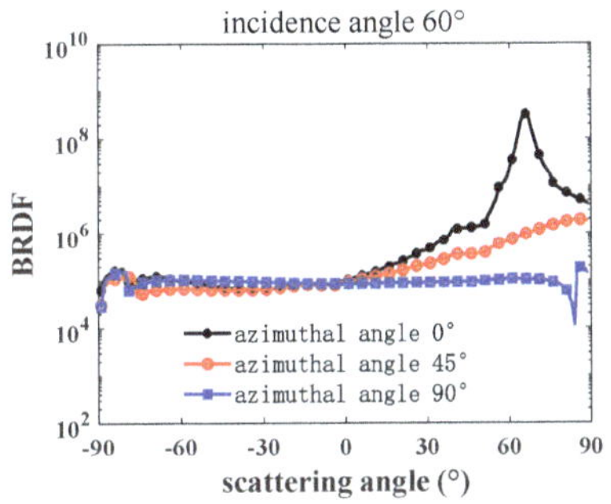

(**d**) Incidence angle 60°

Figure 12. Variation curves of white-light BRDF with scattering angle for oxygen-free copper at different incidence and azimuth angles.

Comparing and analyzing Figures 12 and 13, when the incident angle θ_i = 0°, the scattering intensity observed at azimuth angles of 0° and 45° decreases with the increase in the scattering angle, with the peak at θ_s = 0°, i.e., the strongest scattering of oxygen-free copper in the direction perpendicular to the surface of the material, while the scattering intensity observed at an azimuth angle of 90° is larger than that at other azimuths and is the smallest at a scattering angle of 0° (i.e., perpendicular to the surface of the material), which may be due to the uneven material surface at the measurement point. As with brass and aerospace aluminum surfaces, the BRDF values and scattering intensities at different azimuths are not the same, so it can be seen that the measured oxygen-free copper surface is also anisotropic.

When $\theta_i \neq 0°$ and ϕ_s = 0°, the oxygen-free copper surface exhibits stronger forward scattering than backward scattering, and the BRDF and scattering intensity reaches the peak in the forward scattering direction $\theta_s \approx \theta_i$, which exhibits a very obvious high scattering intensity specular reflection. In contrast, the BRDF and scattering intensity of the brass and aerospace aluminum surfaces gradually increase with the increase in the scattering angle in the forward scattering direction, peak in the specular reflection direction, and then decrease with the further increase in the scattering angle; whereas, except for the high scattering intensity of the oxygen-free copper surfaces in the forward scattering $\theta_s \approx \theta_i$, the scattering intensities of the oxygen-free copper surfaces in other scattering angles are obviously lower than those of the brass and aerospace aluminum surfaces, which suggests that the oxygen-free copper surface has a lower surface scattering intensity. In addition, the highest forward scattering intensity is observed at an azimuth angle of 0°, and the BRDF and scattering intensity in the specular reflection direction increase significantly with the increase in the incidence angle, i.e., the larger the incidence angle is, the oxygen-free copper surfaces have significantly more high-light characteristics in the specular reflection direction.

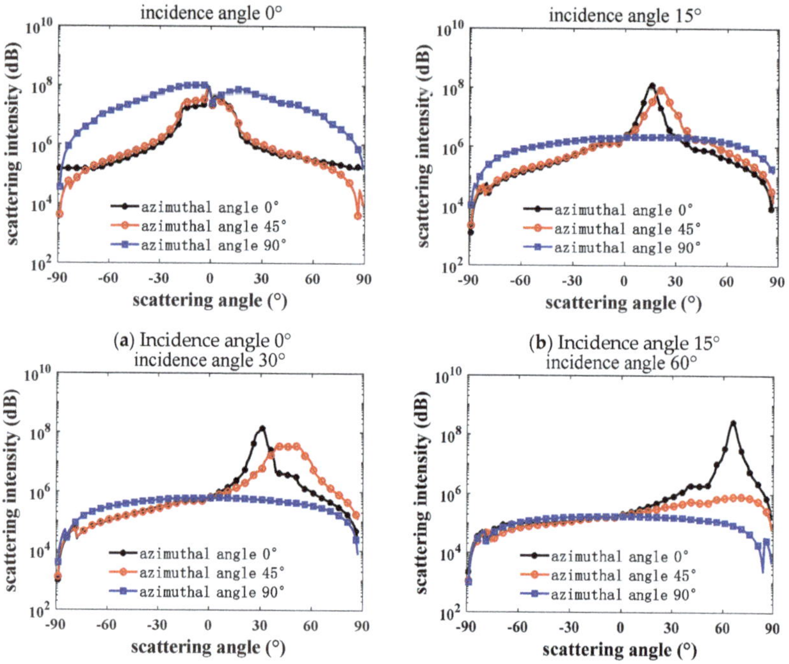

(**a**) Incidence angle 0° (**b**) Incidence angle 15°

(**c**) Incidence angle 30° (**d**) Incidence angle 60°

Figure 13. Variation curves of white light scattering intensity with scattering angle for oxygen-free copper at different incidence and azimuth angles.

When $\theta_i \neq 0°$ and $\phi_s = 45°$, forward scattering is larger than backward scattering. The peaks of the BRDF and scattering intensities occur in directions larger than the specular reflection angle.

When $\theta_i \neq 0°$ and $\phi_s = 90°$, the variation in the BRDF and scattering intensity with the scattering angle is flatter than that on other azimuthal angles, i.e., the scattering light distribution of the material is more homogeneous when viewed at a 90° azimuthal angle. When the incident angle $\theta_i \geq 30°$, the scattering intensity at a 90° azimuth angle is smaller in forward scattering than at other azimuth angles. When $0° < \theta_i < 45°$, the BRDF and the scattering intensity change more gently at smaller scattering angles and decrease with the increase in the scattering angle at larger scattering angles. When $45° < \theta_i < 90°$, the scattering angle is in the range of 84–90°, the scattering intensity increases, and the BRDF becomes significantly larger, i.e., when observed at an azimuthal angle of 90°, the surface of oxygen-free copper appears to be highlighted in this range. Unlike brass and aerospace aluminum surfaces, the angle of incidence has a greater effect on the scattering characteristics of oxygen-free copper materials. Figure 14 illustrates the surface scattering of the oxygen-free copper surface at $\theta_i = 30°$ and $\theta_i = 60°$.

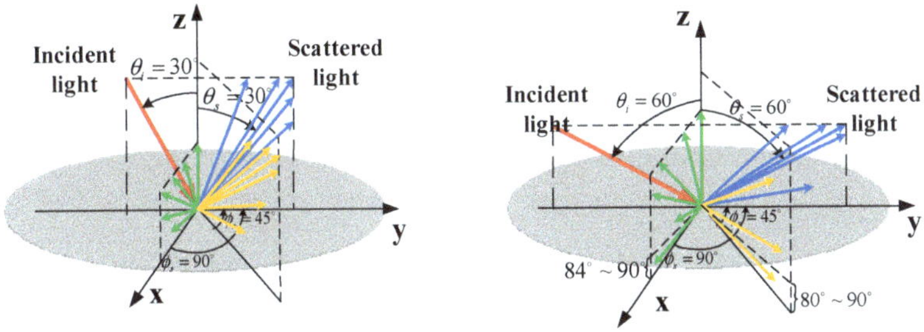

Figure 14. Schematic of surface scattering from oxygen-free copper surface at $\theta_i = 30°$ and $\theta_i = 60°$.

Figures 15 and 16 show the variation curves of the white-light BRDF and scattering intensity with the scattering angle for red copper at different incidence and azimuth angles.

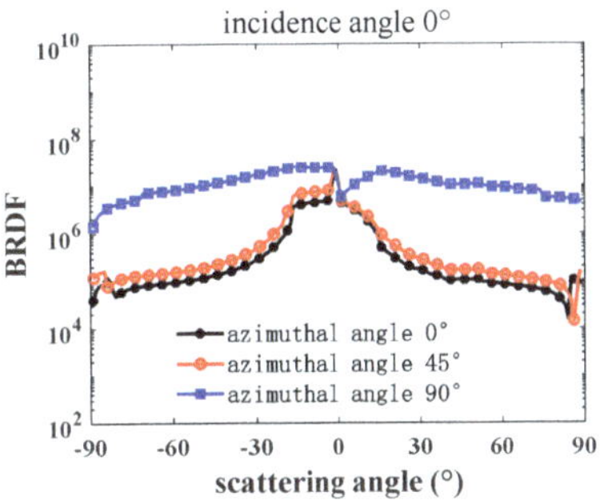

(**a**) Incidence angle 0°

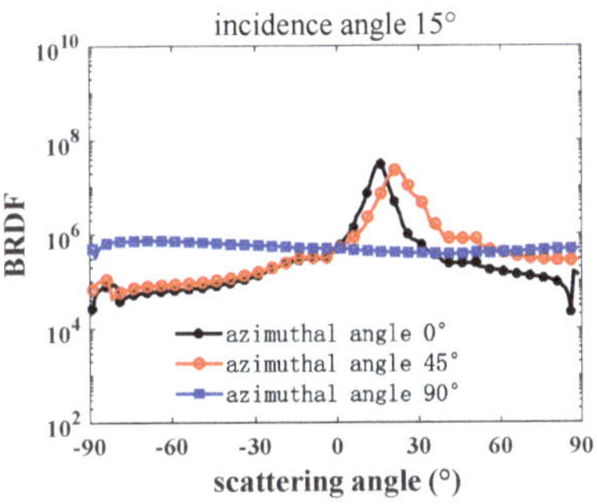

(**b**) Incidence angle 15°

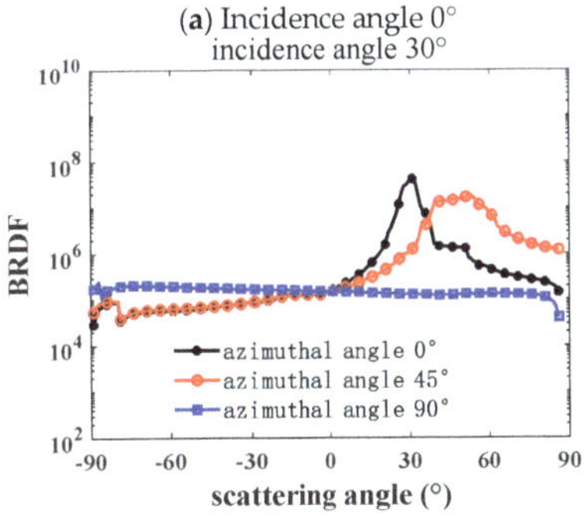

(**c**) Incidence angle 30°

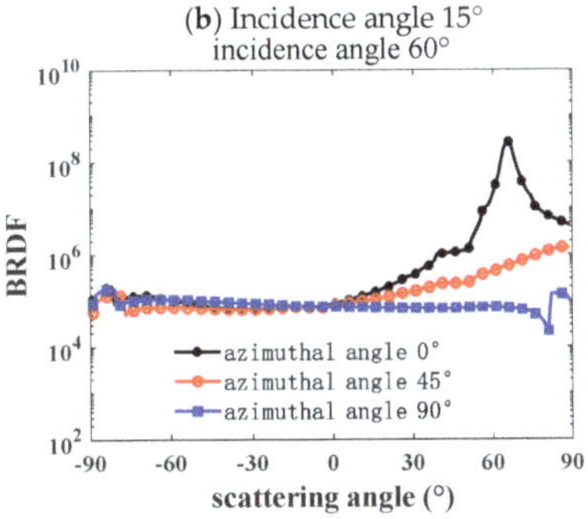

(**d**) Incidence angle 60°

Figure 15. Variation curves of white-light BRDF with scattering angle for red copper at different incidence and azimuth angles.

Comparing and analyzing Figures 15 and 16, the BRDF and scattering intensity of the red copper surface with the incidence angle, scattering angle, azimuthal change rule, and the oxygen-free copper surface are very similar; only the value of the BRDF and scattering intensity is slightly lower than that of the oxygen-free copper surface, which indicates that it has a lower surface scattering intensity, which can be obviously reflected from the values of the two surfaces' roughness, i.e., the roughness of the red copper surface is higher than the roughness of the oxygen-free copper surface, which makes it more suitable for some applications with high requirements on optical surfaces. This is also evident from the roughness of both surfaces, which indicates that the red copper surface has a higher average roughness than the oxygen-free copper surface. This makes the red copper surface more suitable for applications that have higher requirements for optical surfaces.

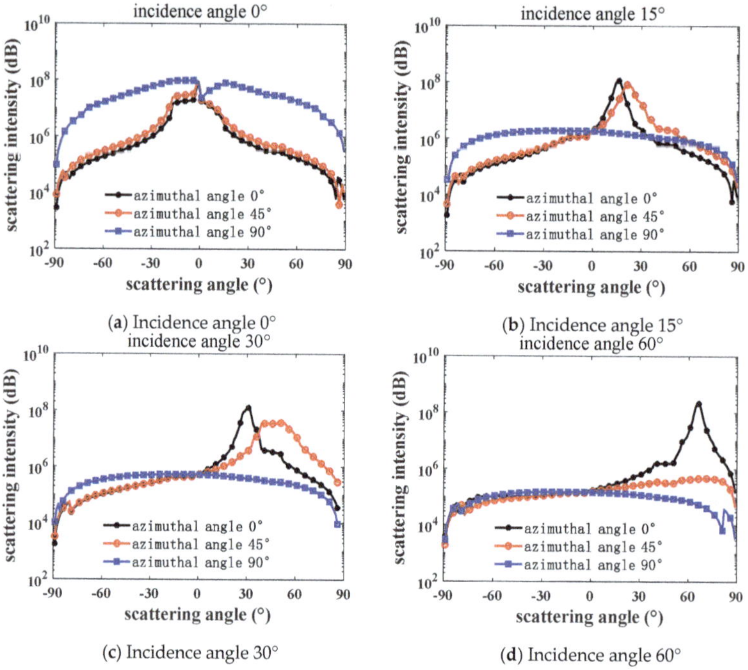

(**a**) Incidence angle 0°

(**b**) Incidence angle 15°

(**c**) Incidence angle 30°

(**d**) Incidence angle 60°

Figure 16. Variation curves of white light scattering intensity with scattering angle for red copper at different incidence and azimuth angles.

A comparison of the surface roughness of the four typical metal materials and the variation curves of the white-light BRDF and the scattering intensity with the scattering angle at different incidence and azimuth angles reveals that, for materials with similar surface roughness values, the oxygen-free copper and red copper surfaces exhibit lower surface scattering intensities and clearer specular reflectance characteristics compared to brass and aerospace aluminum surfaces, which display a pronounced range of high scattering intensities and specular reflections. The specific properties are influenced by the material preparation and treatment process. Furthermore, the incidence angle exerts a more pronounced influence on the scattering characteristics of the oxygen-free copper and red copper surfaces, particularly when the incidence angle exceeds 45°. In such instances, the BRDF and scattering intensity exhibit a pronounced increase at larger scattering angles.

Through the experimental results of this paper, it can be concluded that according to the aforementioned laws, in roughly the same roughness conditions, the four materials according to their own characteristics of the reflection of white light have different performances.

The brass surface can effectively inhibit specular highlights and directional range reflections; the aerospace aluminum surface in the large scattering angle has excellent highlight reflection characteristics; and the red copper surface roughness is slightly higher than the surface roughness of the oxygen-free copper, so the red copper is more suitable for applications in environments with higher requirements for optical surfaces. Under the premise that the surface roughness is roughly the same, the brass and aerospace aluminum surfaces show obvious high scattering intensities and specular reflection ranges, while the oxygen-free copper and red copper surfaces have lower surface scattering intensities and clearer specular reflection characteristics.

Based on the above data, the following analysis is presented.

The brass surface has obvious specular reflection characteristics, manifested in the direction perpendicular to the reflective surface and an azimuth angle of 0° when forward scattering in the direction of the specular reflection has the strongest scattering. At the same time, at a scattering angle of 50–75°, it also has strong scattering, which is appropriate for applications that require specular highlights and directional range reflections.

Aerospace aluminum surfaces also have the strongest scattering in the forward scattering specular reflection direction perpendicular to the direction of the reflecting surface at an azimuth angle of 0° and have better specular high-light performance over a wide range of scattering angles. Aerospace aluminum is suitable for applications that require strong scattering over a wide range of scattering angles.

The oxygen-free copper and red copper surfaces have lower surface scattering strengths and clearer specular reflection properties than the first two materials. The incident angle has a larger effect on its scattering properties, which are significantly enhanced at large scattering angles. They are suitable for applications that require more detailed optical surfaces.

As a result, for different types of optical scenarios, these metallic surfaces exhibit their own unique scattering properties at different incidence, azimuthal, and scattering angles. Depending on their properties, suitable metallic surfaces can be selected to meet specific application requirements, such as specular reflection, directional range reflection, or applications requiring less surface scattering intensity. In addition, material preparation and treatment processes also have an impact on these scattering properties, so these factors need to be taken into account to select the most suitable metallic materials for practical applications.

5. Conclusions

This paper describes the differences and similarities in the scattering characteristics of different metal surfaces by comparing and analyzing the white-light BRDF and the variation in the scattering intensity with scattering angle for brass, aerospace aluminum, oxygen-free copper, and red copper surfaces at different incidence and azimuthal angles and draws the following conclusions.

The brass and aerospace aluminum surfaces exhibit better specular highlighting properties with a higher BRDF and scattering intensity over a wide range of scattering angles. They are appropriate for applications that require specular highlights and directional range reflections.

The oxygen-free copper and red copper surfaces exhibit lower surface scattering intensities. They are appropriate for applications that require sharper specular reflection properties.

The incidence angle has a more significant effect on the BRDF and scattering intensity for the oxygen-free copper and red copper surfaces, which have stronger BRDF and scattering intensity values at large scattering angles than the brass and aerospace aluminum surfaces.

These metal surfaces exhibit their own unique scattering characteristics at different angles of incidence, azimuth, and scattering and are suitable for different types of optical scenarios. According to their characteristics, suitable metal surfaces can be selected to meet specific application requirements, such as specular reflection, directional range reflection, or applications with lower requirements for surface scattering intensity, providing a more comprehensive reference and guidance for the optimization of the material surface processing process and the selection and application of materials for practical engineering applications.

Author Contributions: Conceptualization, Z.Y. and L.G.; data curation, S.Z.; formal analysis, S.Z. and L.W.; funding acquisition, Z.Y. and L.G.; investigation, Z.Y. and Q.X.; methodology, L.W. and L.G.; project administration, Z.Y.; resources, Z.Y. and Q.X.; software, S.Z., Y.L. and L.W.; supervision, Z.Y.; validation, Z.Y.; visualization, S.Z. and L.W.; writing—original draft, Z.Y. and S.Z.; writing—review and editing, L.W. and L.G. All authors have read and agreed to the published version of the manuscript.

Funding: This research was funded by National Natural Science Foundation of China (No. 62201566), the Key Scientific Research Plan of the Education Department of Shaanxi (No. 23JY035), the Youth Innovation Team of Shaanxi Universities (No. K20220184), and the Natural Science Foundation of Shaanxi Province (No. 2024JC-YBMS-523).

Institutional Review Board Statement: Not applicable.

Informed Consent Statement: Not applicable.

Data Availability Statement: Data underlying the results presented in this paper are not publicly available at this time but may be obtained from the authors upon reasonable request.

Acknowledgments: The authors sincerely appreciate all financial and technical support.

Conflicts of Interest: Author Qiang Xu was employed by the company Hubei Huazhong Changjiang OP Toelectronic Technoldgy Co., Ltd. The remaining authors declare that the research was conducted in the absence of any commercial or financial relationships that could be construed as a potential conflict of interest.

References

1. Ishibashi, T.; Horibe, K. Studies on the selective absorption surface on stainless steel. *J. Sun Mankind's Future Source Energy* **19**2, 849–854.
2. Weaver, J.H.; Krafka, C.; Lynch, D.W.; Koch, E.E. Optical properties of metals. *J. Appl. Opt.* **1981**, *20*, 716–719. [CrossRef] [PubMed]
3. Nayar, S.K.; Ikeuchi, K.; Kanade, T. Surface reflection: Physical and geometrical perspectives. *J. IEEE Trans. Pattern Anal. Ma Intell.* **1991**, *7*, 13. [CrossRef]
4. Ward, G. Measuring and modeling anisotropic reflection. *Comput. Graph.* **1992**, *26*, 265–272. [CrossRef]
5. Li, H.; Torrance, K.E. An experimental study of the correlation between surface roughness and light scattering for rough meta surfaces. In Proceedings of the Conference on Advanced Characterization Techniques for Optics Semiconductors, and Nanote nologies II, San Diego, CA, USA, 31 August 2005.
6. Ngan, A.; Durand, F.; Matusik, W. Experimental Analysis of BRDF Models. sucesores de Hernando. *Render. Tech.* **2005**, *200*
7. Xie, M.; Xie, X.; Zhao, H.; Wang, X.; Tan, H. Study on Comparison Experiment of the Bidirectional Reflectance Distribut Function (BRDF) of the Surface of Exterior Decorative Materials for Buildings Under Dry and Wet Condition. In Proceeding the International Heat Transfer Conference, Washington, DC, USA, 8–13 August 2010.
8. Zhang, W.; Zhao, J.M.; Liu, L. Experimental study of the effective BRDF of a copper foam sheet. In Proceedings of the International Symposium on Radiative Transfer, Kusadasi, Turkey, 2–8 June 2013.
9. Vitali, L.; Fustinoni, D.; Gramazio, P.; Niro, A. Spectral BRDF measurements of metallic samples for laser processing applicati *J. Phys. Conf. Series* **2015**, *655*, 012064. [CrossRef]
10. Renhorn, I.G.E.; Boreman, G.D. Analytical fitting model for rough-surface BRDF. *J. Opt. Express* **2008**, *16*, 12892–12898. [Cr Ref]
11. Ďurikovič, R.; Mihálik, A. Modeling the BRDF from spectral reflectance measurements of metallic surfaces. *J. Appl. Surf.* **2014**, *312*, 87–90. [CrossRef]

Cui, J.; Li, X.; Wang, H. Dependence of backscattering characteristics on incident angle and temperature of metal surfaces. *J. Infrared* **2004**, *12*, 1–4.
Wang, M.; Yan, W.; Li, Z.; Lai, J.; Wang, C.; Ji, Y. An optimal modeling of Bidirectional Reflectance Distribution Functionfor coating metals in 1550 nm. *J. Laser Optoelectron. Prog.* **2022**, *59*, 233–238.
Ma, W.; Liu, Y.; Liu, Y. The measurement and modeling investigation on the spectral polarized BRDF of brass. *J. Opt. Rev.* **2020**, *27*, 380–390. [CrossRef]
Zou, Y.; Zhang, L.; Zhang, J.; Li, B.; Lv, X. Developmental trends in the application and measurement of the Bidirectional Reflection Distribution Function. *J. Sens.* **2022**, *22*, 1739. [CrossRef] [PubMed]
Chen, Z.; Liu, Y.; Liu, J.; Yuan, Z.; Liu, Y. The measurement and modeling investigation on the BRDF of brass under variable temperature. *J. Infrared Phys. Technol.* **2022**, *128*, 104505. [CrossRef]
Yupei, M.; Jiaying, C.; Xiaojie, Z.; Zewei, C.; Xiaoli, L. Modeling and Inverse Rendering of Polarimetric Bidirectional Reflectance Distribution Function for Metal Surfaces. *Acta Opt. Sin.* **2024**, *44*, 195–208.
Xie, Y.; Tan, J.; Jing, L.; Zhang, W.; Lai, Q. Investigating directional reflection characteristics of anisotropic machined surfaces using a self-designed scatterometer. *Appl. Opt.* **2019**, *58*, 7970–7980. [CrossRef] [PubMed]
Deng, Z.-L.; Zhang, S.; Wang, G.P. A facile grating approach towards broadband, wide-angle and high-efficiency holographic metasurfaces. *Nanoscale* **2016**, *8*, 1588–1594. [CrossRef]
Aalizadeh, M.; Serebryannikov, A.E.; Ozbay, E.; Vandenbosch, G.A.E. A simple Mie-resonator based meta-array with diverse deflection scenarios enabling multifunctional operation at near-infrared. *Nanophotonics* **2020**, *9*, 4589–4600. [CrossRef]
Margall, F.; Lemineur, G.; Pacanowski, R.; Mavromatis, S.; Coiro, E.; Le Goff, A.; Ceolato, R. Supercontinuum laser-based gonio-scatterometer for in and out-of-plane spectral BRDF measurements. *J. Opt. Express* **2024**, *32*, 25581–25597. [CrossRef]
Li, Y.; Dong, S.; Yu, Q.; Xie, M.; Liu, Z.; Ma, Z. Numerically modelling the reflectance of a rough surface covered with diesel fuel based on bidirectional reflectance distribution function. *J. Opt. Express* **2021**, *29*, 37555–37564. [CrossRef]
Lai, Q.; Liu, B.; Zhao, J.; Zhao, Z.; Tan, J. BRDF characteristics of different textured fabrics in visible and near-infrared band. *J. Opt. Express* **2020**, *28*, 3561–3575. [CrossRef]
Sun, L.; Tang, S.; Zhao, F. An algorithm of computing 3D geometric attenuation factor. *J. Opt. Express* **2019**, *27*, 2056–2073. [CrossRef] [PubMed]
Guo, F.; Zhu, J.; Huang, L.; Li, H.; Deng, J.; Zhang, X.; Wang, K.; Liu, H.; Hou, X. A Modified BRDF Model Based on Cauchy-Lorentz Distribution Theory for Metal and Coating Materials. *J. Photonics.* **2023**, *10*, 773. [CrossRef]
Ferrero, A.; Campos, J. Angular and Spectral Bandwidth Considerations in BRDF Measurements of Interference and Diffraction-Based Coatings. *Coatings* **2020**, *10*, 1128. [CrossRef]
Hou, Q.; Wang, Z.; Su, J.; Tan, F. Measurement of Equivalent BRDF on the Surface of Solar Panel with Periodic Structure. *Coatings* **2019**, *9*, 193. [CrossRef]
Wessels, A.; Christen, L.; Callies, A.; Kroyer, T.; Höhn, O.; Bläsi, B. Modeling of thin-film interference filters on structured substrates: Microfacet-based BSDF versus ray tracing. *J. Opt. Express* **2023**, *31*, 20102–20111. [CrossRef]
González-Leal, J.M.; Gallero, E.; Nuñez, A.; Almagro, J.F. Radiometric analysis of haze in bright-annealed AISI 430 ferritic stainless steel. *J. Appl. Opt.* **2022**, *61*, 2155–2164. [CrossRef]
Butler, S.D.; Marciniak, M.A. Utilization and efficient computation of polarization factor Q for fast, accurate BRDF modeling. *J. Opt. Express* **2022**, *30*, 5803–5816. [CrossRef] [PubMed]
Wessels, A.; Callies, A.; Bläsi, B.; Kroyer, T.; Höhn, O. Modeling the optical properties of Morpho-inspired thin-film interference filters on structured surfaces. *J. Opt. Express* **2022**, *30*, 14586–14599. [CrossRef] [PubMed]
Small, T.V.; Butler, S.D.; Marciniak, M.A. Solar cell BRDF measurement and modeling with out-of-plane data. *J. Opt. Express* **2021**, *29*, 35501–35515. [CrossRef]
Heitz, E. Understanding the Masking-Shadowing Function in Microfacet-Based BRDFs. *J. Comput. Graph. Tech.* **2014**, *2*, 32–91.
Renhorn, I.G.E.; Hallberg, T.; Bergström, D.; Boreman, G.D. Four-parameter model for polarization-resolved rough-surface BRDF. *J. Opt. Express* **2011**, *19*, 1027–1036. [CrossRef] [PubMed]
Butler, S.D.; Nauyoks, S.E.; Marciniak, M.A. Comparison of microfacet BRDF model to modified Beckmann-Kirchhoff BRDF model for rough and smooth surfaces. *J. Opt. Express* **2015**, *23*, 29100–29112. [CrossRef] [PubMed]
Shuang, D. Research on the reflection characteristics of oil films on rough sea surfaces based on BRDF. *Dalian Marit. Univ. Doc. Diss.* **2025**, 19–21.

MDPI AG
Grosspeteranlage 5
4052 Basel
Switzerland
Tel.: +41 61 683 77 34

Coatings Editorial Office
E-mail: coatings@mdpi.com
www.mdpi.com/journal/coatings

www.ingramcontent.com/pod-product-compliance
Lightning Source LLC
LaVergne TN
LVHW070446061126
851858LV00011B/172